No. 529
$7.95

Handbook Of Magnetic Recording

By Finn Jorgensen

TAB BOOKS
BLUE RIDGE SUMMIT, PA. 17214

FIRST EDITION

FIRST PRINTING — JULY 1970

Printed in the United States
of America

Library of Congress Card Number: 77-114711

Contents

Introduction

Magnetic recording has come to play a major role in our modern society. Speech, music, telemetry data, bookkeeping, computations, and live pictures are today recorded, stored, and replayed on magnetic tapes or discs.

The success of magnetic recording is found in its convenience of use, low cost, and reusability of tapes. It is quite paradoxical, however, that many tapes are recorded only once. When the voice of a family member or friend far remote is heard, or a favorite music number finally recorded, one is quite hesitant to erase it. This also applies in the recording of scientific data, where the tapes normally are stored in libraries.

Magnetic recording was invented by Valdemar Poulsen in 1898 and the first recording machine was completed in 1899. It won a Grand Prix in 1900 in Paris at the World Exhibition. This unique recorder/reproducer was called the Telegraphone and consisted of a brass drum, spirally wound with piano wire which was the recording medium. The magnetic recording and playback head was mounted on a sled on top of the drum and it progressed to the right as the drum revolved. At the end, it lifted from the drum and moved to the left by means of a lead screw. Then, it was ready for a repeat pass along the drum.

A company was formed in the U.S.A. to manufacture and sell the Telegraphone, but its success was shortlived since the maximum playing time was one minute. Edison's phonograph was superior, having a lower cost and longer playing time.

Magnetic recording lay essentially dormant until the Second World War, during which time it was employed by the German forces. They succeeded in replacing the piano wire with a flat plastic ribbon, containing magnetic particles and wound onto a reel. This produced much longer playing time and the recordings were made with an additional ingredient: AC bias. The recording quality was improved some twenty decibels by the addition of the AC bias, which now is used in all

I-1. Valdemar Poulsen's first wire recorder, the telegraphone.

analog recorders (but not in digital recorders). It is paradoxical that AC bias dates back to 1927, when a patent was granted to W. L. Carlson and G. W. Carpenter of the U. S. Navy. Few know that Valdemar Poulsen in 1902 had another invention in his laboratory, a 100-kHz generator. It would have served very well as a bias generator for his Telegraphone.

After World War II, a few magnetophones were brought to the United States and a new evolution of magnetic recording started. A large number of wire recorders, using AC bias, were manufactured and sold, and when coated magnetic tapes were introduced by the 3M Company in 1947, the recording industry developed rapidly. The first television recorder was designed and built in 1951 by Bing Crosby Enterprises, using a tape speed of 360 IPS (inches per second) and a tape width of one inch. The high speed was necessary to achieve a reasonable frequency response. The video signal was further split into fourteen channels on the tape and recombined during playback. Later on, in 1956, Ampex developed a television recorder, using a two-inch wide tape with a longitudinal speed of 15 IPS and a transverse rotating head assembly, whereby the actual tape-to-head speed was in the order of 1500 IPS.

Magnetic recording has since then migrated into every phase of life, enough to fill a rather voluminous book on the applications and the various recorder types. But they are all familiar in principle, and this book is devoted to an explanation of these principles. This will give the reader an understanding of the functions in magnetic recording and will be of help to those involved in the design, use, and maintenance of recorders. There are no detailed descriptions of individual recorders, although some units are shown and explained as typical examples of their types. Many maintenance problems are common, such as debris on magnetic heads, magnetized heads, or incorrect operation of recorders, and these topics are discussed. The contents has been divided into four sections, so as to better serve as a textbook and a handbook:

Fundamental explanation of magnetic recording and the tape recorder............................Chapters 1 and 2

Detail theory of the tape transport, magnetic heads, tapes, amplifiers, and equalization.............Chapters 5, 6, and 7

Selection of tapes and accessories, applications, proper use of the recorder, care and maintenance, and specialized techniques........................Chapters 8, 9, 10, and 11

Tables, formulas, standards and measurement techniques....................Chapter 12

Chapters 3, 4 and 5 (and 12) can be bypassed by the reader who seeks only information about the basic principles for a better understanding of his recorder and how to best use it. The book in its entirety should be of value for the students, technicians, and the engineers involved with magnetic recording. Supplementary readings are suggested in the selected references at the end of each chapter.

The author wishes to thank many friends and associates in the recording industry for valuable suggestions and assistance during the making of this book. Also thanks to Mrs. Sue McCandless and Mr. Art Carson for great help with the manuscript.

Chapter 1

Magnetic Recording & Playback

The basic principles of magnetic recording and playback are explained in this chapter and the following chapter describes the recorder itself with its functions of recording, playback and tape movement.

The principles of **magnetic recording** are based on the physics of magnetism, a phenomenon which relates to certain metallic materials. Magnetization of materials occurs when they are placed in a magnetic field. If the material is in the group of so-called "hard" magnetic materials, it will hold its magnetization after it has been moved away from the exciting source.

Fig. 1-1 is a simplified diagram of a recorder. An incoming sound wave is picked up by a microphone (1) and amplified (2) into a recording current, Is (3), which flows through the winding in the record head. The record head has a "soft" magnetic core (so magnetization is not retained) with an air gap in front. The current, Is, produces magnetic field lines that diverge from the air gap (4) and penetrate the tape, moving past the record head from the supply reel (5). The tape itself is a plastic ribbon coated with a "hard" magnetic material which maintains its magnetization after it has passed the gap in the record head.

The tape passes over the playback head which, like the record head, is a ring core with a front gap. The magnetic field lines (flux) from the recorded tape permeate the core and produce an induced voltage, E (6), across the winding. This voltage, after suitable amplification (7), reproduces the original sound through a speaker (8).

This fundamental record and playback process works, but its use is limited by the accompanying poor fidelity in music and data recording. It is used only in computer applications, where the sole criteria for performance is the presence or absence of a signal. In high-fidelity music recordings and in instrumentation recordings, an additional (bias) current, Ib (9), is added to the record current flowing through the record head winding. The bias is a high-frequency current that

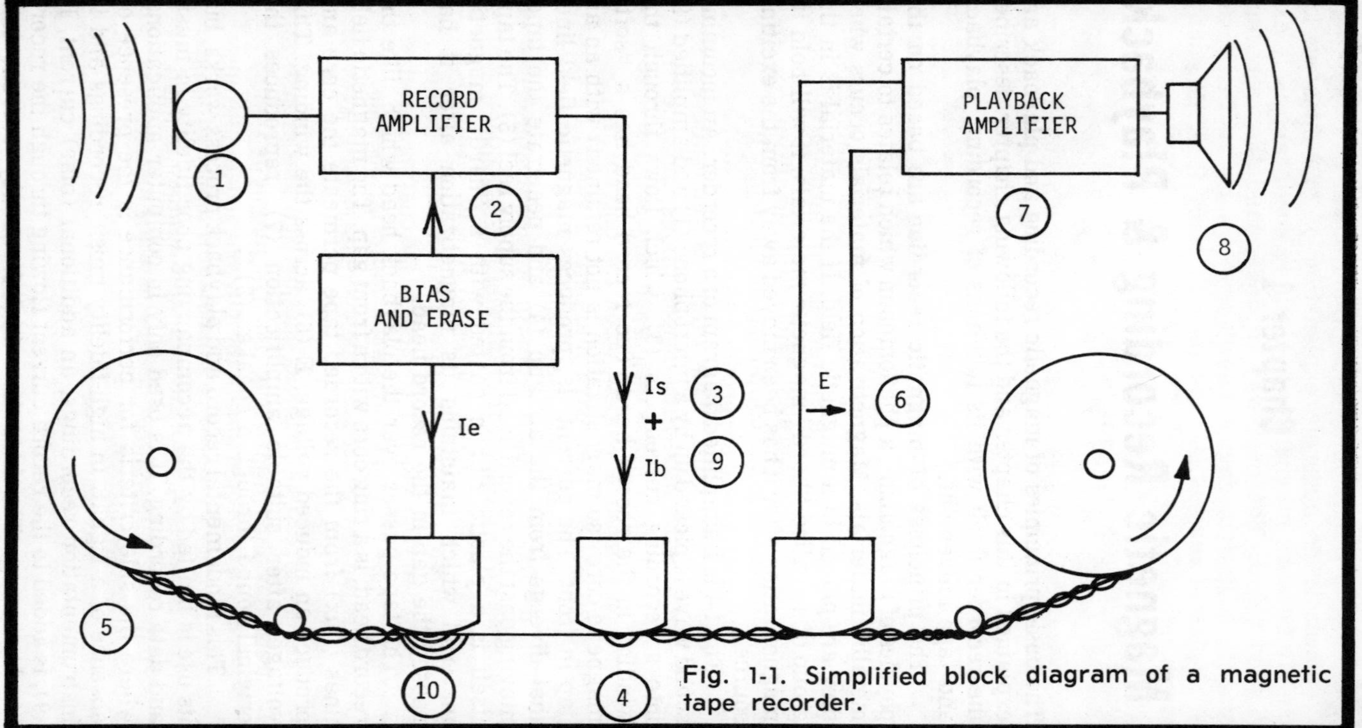

Fig. 1-1. Simplified block diagram of a magnetic tape recorder.

provides a great improvement in recording fiedlity and a simultaneous reduction of background noise.

There is no accurate explanation of how bias works mainly because magnetism in itself is nonlinear, as is explained below. The best way to understand the recording process is first to gain a knowledge of some of the basic principles of magnetism and then to apply it to magnetic recording and playback.

INTRODUCTION TO MAGNETISM

Certain metallic materials have the ability to become magnetized and can basically be divided into three groups:

Non-magnetic: Copper, aluminum, brass

Soft magnetic: Mu-metal, permalloy, transformer laminations

Hard magnetic: Alnico, iron oxide (tape coatings)

The difference between these materials is best illustrated by a simple experiment where a sample of each material is inserted into a coil. When a current is sent through the coil, a magnetic field is generated, and with the presence of a magnetic material in the coil forms an electromagnet. A non-magnetic material will not produce any change in the coil field but the other two will. (This effect is remembered from physics, where the attraction of iron filings indicate the increase in field lines.) The difference between soft and hard magnetic materials is evident when the current through the coil is switched off. The hard magnetic material retains its magnetism, while the soft material loses it.

The presence of both types of magnetic materials increases the coil's magnetic field strength by a property named permeability, but each differs in its retention of magnetization. Hard magnetic material keeps its magnetization; therefore, it is used for the coating material on magnetic tapes. The soft magnetic material does not retain magnetization and is used for the core material in recording and playback heads.

The characteristic of being magnetized and of retaining magnetism is illustrated in Fig. 1-2. The magnetic field from a coil increases in proportion to the current flowing through it; therefore, it shows a linear relationship. If a core of magnetic material is inserted into the coil, the number of field lines around the coil will increase. The ratio of the increase over air is called the permeability (μ), which varies from one to several thousand and depends on how well the material conducts magnetic field lines. But there is a limit: As the field strength increases, the magnetic material reaches a

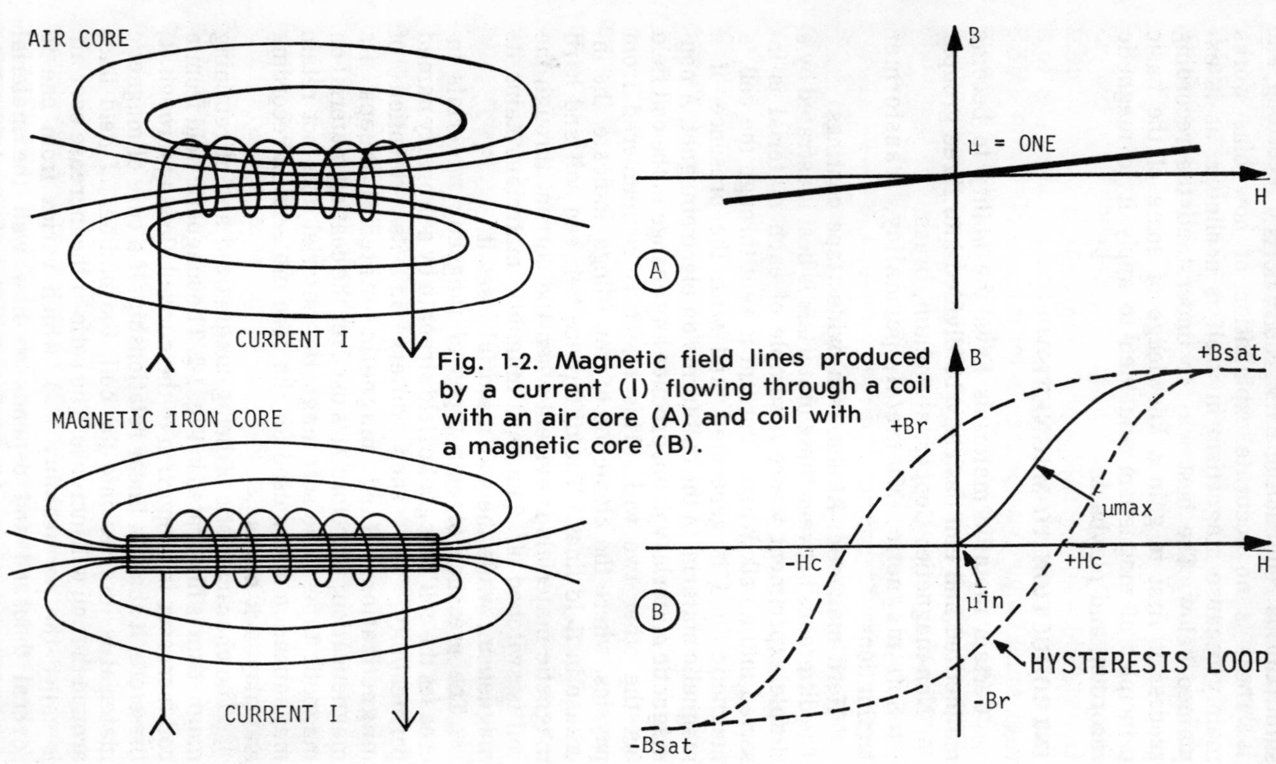

Fig. 1-2. Magnetic field lines produced by a current (I) flowing through a coil with an air core (A) and coil with a magnetic core (B).

saturation point, and the permeability starts to decrease. This effect causes nonlinear behavior of a magnetic recording, and is shown by the solid line in Fig. 1-2B.

The number (density) of field lines within the material is called the **induction, B,** and is related to the **field strength, H,** of the coil by a simple formula:

$$B = \mu \times H$$

Permeability (μ) is not a constant; it depends upon the external field, H, and the magnetic material. It has an initial value (μin; see Fig. 1-2) greater than one and increases to a maximum value (μmax), thereafter it levels off to one. (MKS units are used throughout this book.) Consequently, high induction (desired in magnetic heads) is obtained with a material having a high permeability. The value of μ is equal to one for air and non-magnetic materials, but can be 10,000 or higher for soft magnetic materials such as are found in the cores of recording and playback heads.

The induction in a magnetic material will not return to zero when the external electric field is removed. This property is called the **remanence (Br)** and is an important parameter for magnetic tapes; it indicates how well a tape will keep its magnetization after it passes the recording head. If the direction of the current through the coil (Fig. 1-2B) is reversed, the induction B will decrease from Br to zero and then become negative (as shown by the broken line). And if an alternating current is used, the induction will likewise alternate between positive and negative values. The relation between B and H then will follow what is called a **hysteresis loop,** going through the values of plus Br and minus Br on the vertical axis and plus Hc and minus Hc on the horizontal axis. The field strength Hc is called the **coercive force.** At very high values of field strength, the induction approaches **saturation** named Bsat.

The initial magnetization curve (from an unmagnetized state to saturation, the solid line in Fig. 1-2B) and the hysteresis loop (shown by the broken line) define and describe the characteristics of a magnetic material. The quality of a recording depends upon the linearity between the field strength in the record gap and the magnetization (remanence Br) left on the tape. A doubling of the field strength should result in a doubling of the remanence. Fig. 1-3 illustrates the relation between the field strength H and the remanence Br. This relationship is illustrated as being derived from the remanence values obtained by the increasing values of the field strength as shown in Steps 1 to 8, Fig. 1-3. The remanence from an alternating field can be constructed by a graphic projection from the remanence values in Fig. 1-3

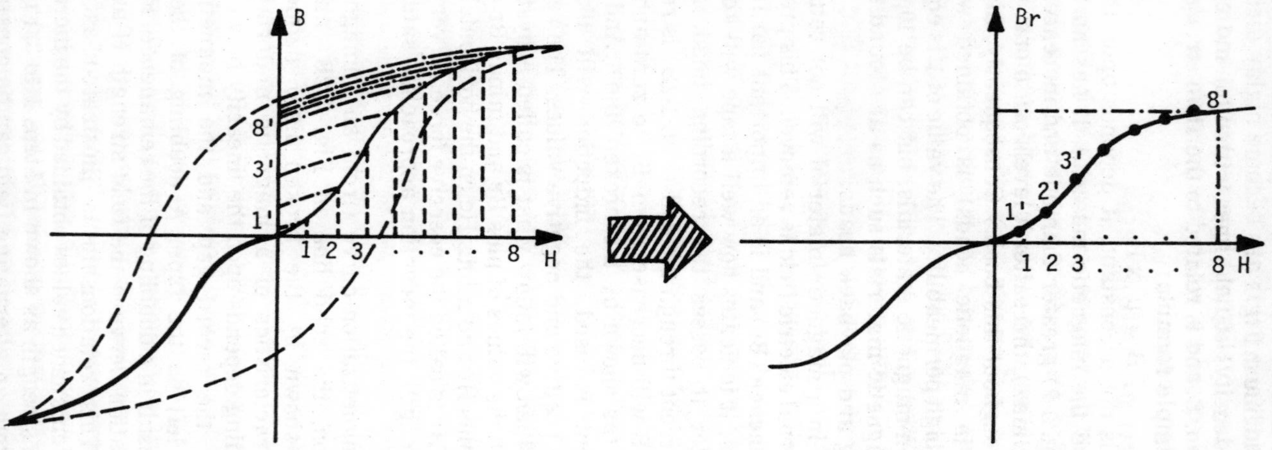

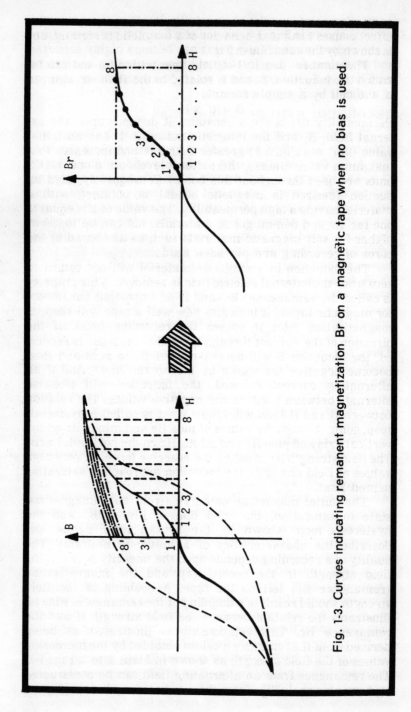

Fig. 1-3. Curves indicating remanent magnetization Br on a magnetic tape when no bias is used.

and it is shown in Fig. 1-4. As you can see, the pure waveform of the record current (and field) has become highly distorted and is useless for any intelligent voice recording and can be used only in digital recorders, where the presence or absence of a signal is of significance.

RECORDING WITH DC AND AC BIAS

The midrange of the remanence curve is quite linear, and if the range A' A (Fig. 1-5) could be avoided by using either A-C or A'-C', the remanence on the tape would be more linear. This was recognized in the early days of magnetic recording and its effects were minimized by offsetting the recording field strength with an additional DC current through the head winding. This is shown in Fig. 1-5A, where the offsetting field Hdc has moved the **working point** to Pa and the resultant remanence of the tape coating has a DC component Bdc and a superimposed cleaner alternating remanence.

A doubling of the remanence is possible if the tape is passed over a permanent magnet before reaching the record head. The magnet saturates the tape coating into the negative region (-Br). As a result, the DC field moves the working point to Pb and a larger linear range is obtained (A''-C'', Fig. 1-5B). While this recording technique is adequate for digital and some dictating recorder uses, it does introduce a major disadvantage in audio and data signal recordings—a high noise level. The magnetic coating on the tape does not have a perfectly uniform magnetic characteristic and this results in flux variations.

The tape coating consists of a plastic binder with small magnetic particles imbedded in it. Each particle contributes to the number of external field lines remanent after recording. If all particles were magnetized in the same direction and uniformly dispersed, the induced voltage in the playback head winding would be zero. But this ideal condition is never achieved. The variations due to the lumping of the particles and their varying sizes produce a noise level which in an audio recorder limits the signal-to-noise ratio to 40 db.

In the late twenties, it was discovered that the addition of a high-frequency signal to a sound or data signal greatly improved the quality of magnetic recording. This technique is now called **AC bias recording** and it has become an important ingredient in all the recorders of today. The action of AC bias is not explained by a simple formula, well defined by physical laws, since the magnetization process of the tape is highly nonlinear. The simplest explanation is best given by studying the effect of AC fields. Fig. 1-6A illustrates how a magnetic material is demagnetized by an alternating field that slowly

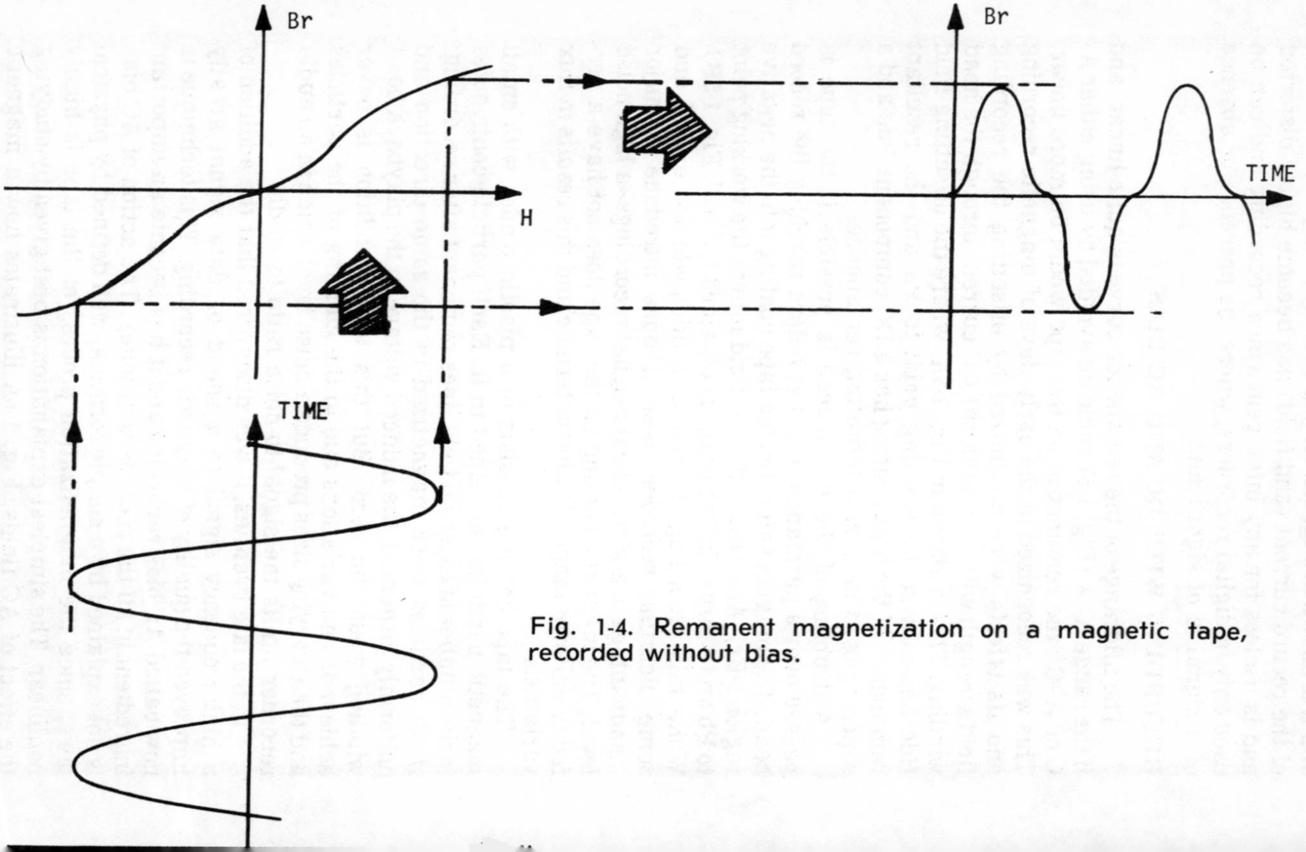

Fig. 1-4. Remanent magnetization on a magnetic tape, recorded without bias.

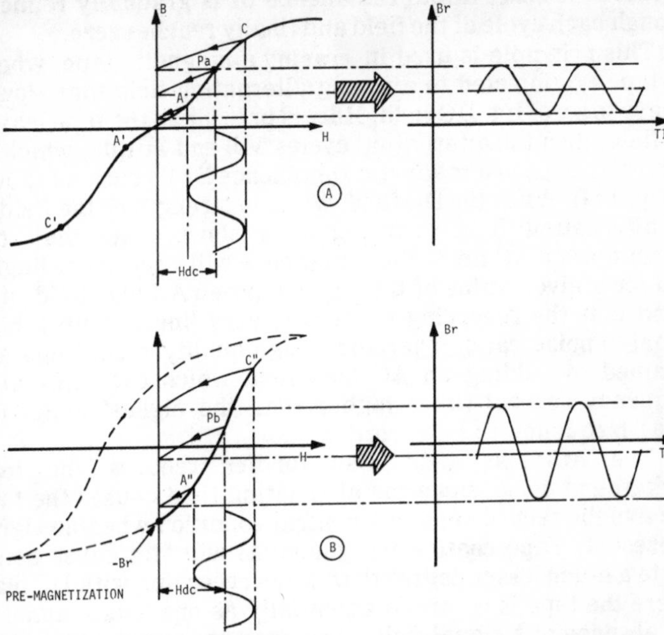

Fig. 1-5. Curves showing effect of DC bias: (A) using a neutral tape, and (B) using a pre-magnetized tape.

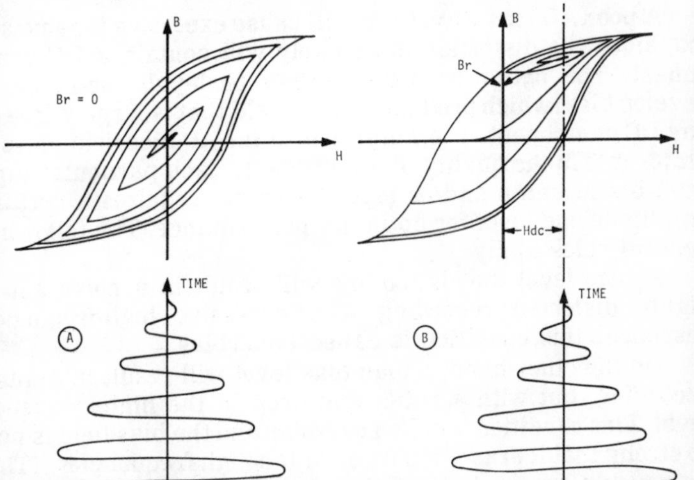

Fig. 1-6. Magnetic remanence Br resulting from a decaying alternating field. (A) alternating field alone; (B) alternating field with superimposed DC field Hdc.

decreases to zero. Initial remanence Br is gradually reduced through each cycle of the field and finally reaches zero.

This principle is used in **erasing** a magnetic tape, where the tape is subjected to a strong alternating field that slowly decays to zero. If a DC field, Hdc, is present as the tape leaves the field, then the alternating cycles will end at Hdc, which in turn will result in a magnetic remanence Br-dc such as shown in Fig. 1-6B. When the DC field varies in strength or the field is an alternating field (varying at a slower rate than the superimposed AC field) the remanence will vary accordingly. And for a given value of the superimposed AC bias field, it is found that the recording process is very linear, with a high signal-to-noise ratio. Therefore, high-quality recordings are obtained by adding an AC bias field which oscillates at a frequency several times higher than the highest audio (or data) frequency to be recorded.

The AC bias field also further reduces the tape background noise, since the alternating field causes the tape to leave the record head in a neutral condition when no signal is present. Tape coating irregularities will still cause noise, but to a much lesser degree than when recording with DC bias, where the tape is recorded essentially as one long magnet in the absence of a signal field. Any coating irregularities show up as a noise output caused by a change in the external tape flux.

A poor AC bias waveform will cause excessive tape noise, too, since the distortion most likely will contain a DC component. The better recorders use a push-pull oscillator to develop bias, which produces an AC field that is largely free of any DC component. The **amplitude of the AC bias** field plays a major role in the quality of a recording. Each particular tape (i.e., brand name and/or tape thickness) requires a certain amplitude and level for optimum performance. The following general rules apply:

A bias level that is too low will result in a noisy and a highly distorted recording with excessive high-frequency response; this condition is called underbias.

On the other hand, a high bias level will result in a quiet recording, but with a noticeable drop in the high-frequency level. This condition is called overbias and the bias field is now so strong that it erases a portion of the high frequencies. (This also would be the case if the magnetic heads were out of alignment or had a buildup of foreign materials; for example, dust or debris from the tape surface. See Chapter 10.) Chapter 6 deals further with the magnetic recording process and AC bias.

PLAYBACK FREQUENCY RESPONSE

When the magnetic tape leaves the record head, it has a permanent magnetic record of the sound or data signal, with flux lines extending from the surface, as shown in Fig. 1-7. These flux lines pass in sequence across the reproduce-head core and induce a voltage, E, which after amplification reproduces the original signal. The number of flux lines is proportional to the recorded signal strength and their duration is inversely proportional to the recorded frequency. Their duration represents a certain **wavelength** on the tape and can be expressed as:

$$\text{Wavelength} \quad \lambda = \frac{\text{Tape speed v}}{\text{Frequency f}}$$

It is a general practice to run tape at slow speeds to reduce the quantity of tape needed and to design magnetic heads and tapes for good short wavelength performance. If a high-frequency response of 15 kHz is desired, as in the reproduction of high-fidelity music, the wavelength should be as short as possible to save tape. If you want to record and reproduce 15 kHz at a tape speed of 3¾ IPS (9cm/sec), the wavelength is:

$$\begin{aligned} \text{Wavelength} &= \frac{3.75}{15{,}000} \text{ inches} \\ &= .25 \text{ mils} \\ &= 6\,\mu \end{aligned}$$

This is an extremely short length—only one quarter of the thickness of a long-play magnetic tape. And the reproduce (playback) head gap must be wider than half of that wavelength, otherwise the high-frequency response will suffer and the fidelity of the recording will be lost. (See Chapter 6.)

The **induced voltage** (E) in the reproduce head winding is very low and requires amplification of 10 to 100,000 times to provide a useful output. This is particularly true in modern recorders with micro-gaps, narrow track widths (four or eight tracks are recorded on a quarter-inch wide tape in some cases), and low tape speeds. These factors complicate the noise problem, a design consideration that is aggravated by the fact that the induced voltage is lowest at low and high frequencies. If all signal frequencies were recorded at the same level we would expect that the corresponding wavelengths on the tape would be of equal strength. But in reality, we find that the flux strength decreases rapidly toward short wavelengths, so the actual flux from the tape follows a pattern as shown in Fig. 1-8.

Thus, the voltage (E) induced in the playback head is not only proportional to the tape flux but also to frequency. This

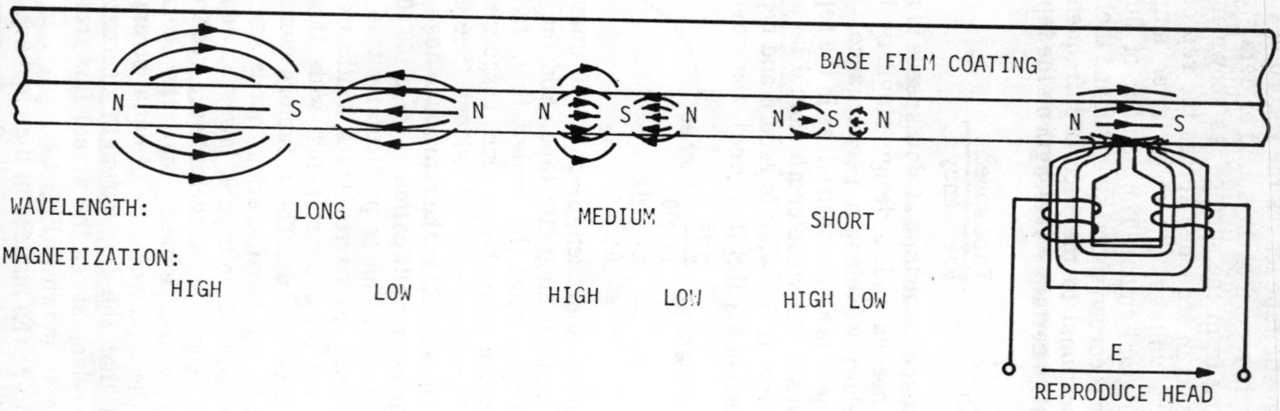

Fig. 1-7. Recorded flux patterns on a magnetic tape.

tends to compensate for the decreasing flux, with a voltage vs frequency curve as shown in Fig. 1-8B. In order to achieve a constant output voltage over the entire frequency range, more amplification must be provided for low and high frequencies, as shown in Fig. 1-8C. This is what is technically referred to as equalization. From the drawing it is also seen that the otherwise flat noise level increases at the low and high ends, which is unfortunate since it emphasizes any amplifier hum and tape hiss.

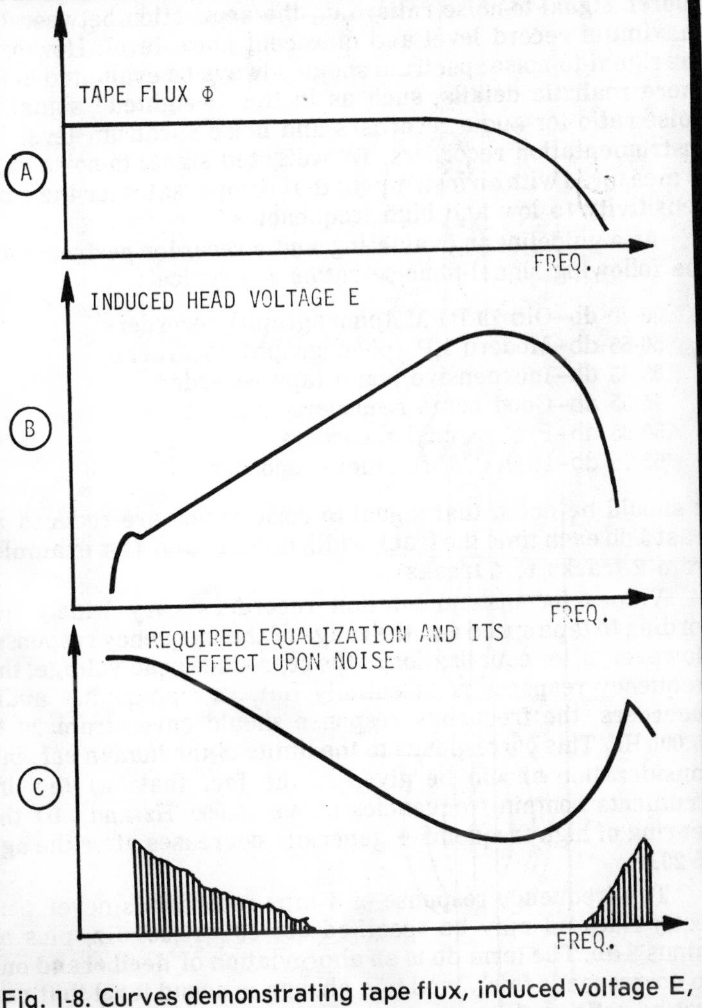

Fig. 1-8. Curves demonstrating tape flux, induced voltage E, and the required equalization.

It is standard practice to boost the record current in these high- and low-frequency regions so that less playback equalization is required. Too much boost may cause an overload condition with subsequent distortion, so the audio industry has established standards for equalization (see Chapters 7 and 12). These standards rely upon the knowledge of sound level vs frequency in music and speech. When this knowledge does not exist, as in research and instrumentation, the boost of the record current cannot be tolerated and instrumentation recorders, therefore, always show an apparent poorer **signal-to-noise ratio**; i.e., the separation between the maximum record level and quiescent noise level. However, the signal-to-noise spectrum should always be evaluated in the more realistic details, such as in the "weighted" signal-to-noise ratio for audio recorders and noise spectrum level for instrumentation recorders. (A weighted signal-to-noise ratio is measured with an instrument that compensates for the ear's sensitivity to low and high frequencies.)

As a guideline in evaluating **audio recorder performance**, the following signal-to-noise ratios are typical:

35-40 db–Old 78 RPM (phonograph) recorders
50-65 db–Modern LP (phonograph) recorders
35-45 db–Inexpensive home tape recorders
45-55 db–Good home recorders (2 tracks)
50-65 db–Professional recorders
65-75 db–High-quality studio recorders

It should be noted that signal-to-noise ratios are reduced at least 3 db each time the track width is cut in half (for example, from 2 tracks to 4 tracks).

Values for instrumentation recorders vary widely according to tape speed and the range of the frequency response. However, after equalization of the reproduce head voltage, the frequency response is essentially flat. In high-quality audio recorders, the frequency response should cover from 20 to 16,000 Hz. This corresponds to the limits of the human ear, but consideration should be given to the fact that: a) few instruments contain frequencies above 10,000 Hz and, b) the hearing of high frequencies generally decreases after the age of 20.

The frequency response of a tape recorder is never perfectly flat and may be specified as: 50 - 16,000 Hz, plus or minus 3 db. The term db is an abbreviation of **decibel** and one db corresponds fairly well to a change in sound level that can just be noticed. As a rule of thumb, we can use the following scale:

Level change:	1 db -	Barely noticed by an expert under ideal conditions.
	3 db -	Noticed under normal listening.
	6 db -	A definite change in sound level.

A specification of plus or minus 3 db corresponds quite well to what the ear can tolerate. It should also be noted that **frequency response** applies to the record and reproduce electronics, using a good grade magnetic tape; it does not include microphones or speakers often built in or supplied with the recorder. And the requirements for the frequency response vary with the applications and program source:

Live recordings (record companies, FM broadcast use)	20-20,000 Hz (15 IPS)
Recordings of FM programs	30-15,000 Hz (7 1/2 IPS)
Recordings of AM programs	50-6,000 Hz (3 3/4 or 1 7/8 IPS)

The most often used tape speeds are indicated in parentheses, IPS being the abbreviation for inches per second. Again, very few instruments have sound components above 10,000 Hz, and a speed of 3 3/4 IPS is in many cases fully adequate for home recordings of FM programs. As the above table indicates, the frequency response does improve with increasing tape speed and it is in fact more logical to speak of how many wavelengths can be recorded per inch of tape. 2,000-5,000 Hz per inch of tape length is common practice for audio recorders and the present limitation is 20,000 Hz per inch. This requires high-quality tapes and good magnetic heads, but it is an improvement of 10 times over the past 20 years. (The frequency-response requirements of video (TV) and instrumentation recorders are discussed in Chapter 11.)

FULL-TRACK, HALF-TRACK, QUARTER-TRACK, AND OTHER STANDARDS

Twenty years ago, all magnetic recordings were made on 1/4" wide tape with the full width of the tape being used for one channel (monaural). Improvements in tape quality came rapidly, especially during the '60s, and the amateurs and high-fidelity enthusiasts took advantage of this improvement and began recording on only half the width of the tape (called half- or two-track recording). After playing one side, the reel is simply turned over for the other half (Fig. 1-9.) By cutting the track width in half, only half the flux is available, but the improved tapes made up for the difference. (Otherwise, the

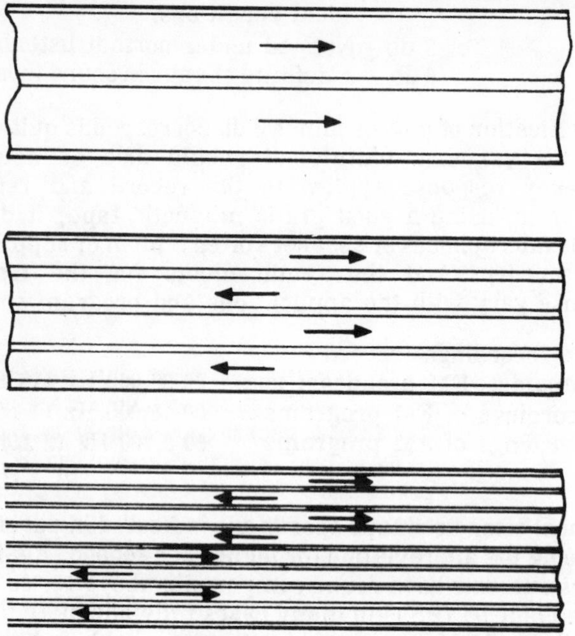

Fig. 1-9. 2-, 4-, and 8-track arrangement on quarter-inch magnetic tape.

signal-to-noise ratio would have decreased by a factor of 3 to 6 db.)

With the advent of stereo (or binaural) recordings, the two tracks were used for the two channels. With further improvements in tape quality, the standards were once again changed to what now is called four- (or quarter-) track tapes, two tracks being used for stereo recording in one direction and the remaining two tracks for the reverse direction. Latest developments have made possible the recording and playback of up to eight tracks on quarter-inch tape, the method used with prerecorded stereo music cartridges. Eight-track recording is primarily used in car installations, where the noise level is somewhat high and the poorer signal-to-noise ratio of the narrow track width (21 mils equal 0.5mm) is not too critical.

Other standards are used in the instrumentation and computer fields where the most common is seven or nine tracks on 1/2" tape for computers and seven tracks in instrumentation recording. Also, the trend in these fields is toward the narrower tracks.

QUALITY VS TRACKWIDTH

The voltage induced in the playback head, as mentioned earlier, requires equalization, which in essence is a boost of the low and high frequencies. The low-frequency boost, of course, tends to emphasize hum and transistor noise, while the high-frequency boost increases the noise from the tape itself and the noise that may be generated in the head, both having the characteristic of hiss. When the track width is reduced, so is the induced voltage. Since the noise is the same, the signal-to-noise ratio obviously drops.

The narrower tracks also increase the possibility of **crosstalk**, since the flux lines from an adjacent track can leak over into the reproduce head. This is particularly true of eight-track recordings on 1/4" tape, but it also can occur in four-track recordings. Crosstalk is best avoided by using carefully designed magnetic heads, proper guidance of the tape past the head, and the use of high-grade tape having a well-controlled width (slitting tolerance).

All home recorders have an **erase head**, in addition to the record and reproduce heads. The erase head is always activated during recording and provides a very strong and slowly decaying AC field that erases the tape. The erase head is mounted before the record head (see No. 10 in Fig. 1-1). The erase head does not erase the tape across its full width but only those tracks being recorded. To completely erase a reel of tape, a bulk degausser is normally used. A degausser is a powerful AC electromagnet which is brought near the tape reel and then slowly removed. This again provides the action illustrated in Fig. 1-6A.

Chapter 2

The Tape Recorder

A magnetic tape recorder is a device for the recording and playback of sound, music or data information. It is essentially constructed in two sections:

1. A tape deck (or transport) that moves the tape past the recording and playback heads. It has the necessary controls for starting and stopping the tape motion.

2. An electronic section that contains the recording and playback amplifiers. This includes an erase/bias oscillator section and power supplies. Most recorders are provided with level indicators to assure proper recording levels.

Fig. 2-1 shows the basic elements in a magnetic recorder/reproducer, with a distinction between the transport and the electronics. The two sections may be combined into one unit or they may be packaged into two separate units, with interconnecting cables. The first concept is used in most home entertainment recorders and in some instrumentation recorders, where portability is required. In studio recorders, where space limitations are less stringent, it is generally found that the tape recorder is designed into separate transport and electronics units. This in turn facilitates both design, fabrication, and maintenance of the recording equipment.

THE TAPE TRANSPORT

Among the variety of recorders manufactured today, there are probably not two with transports that are exactly alike. But the differences in general are reflected only in cost, quality, and style, while the basic functions are common to them all. The art of designing and mass producing recorders has come a long way over the past decade without any startling changes in performance. However, there have been reductions in price and in many cases other features have been added. It is basic to all transports that they move the tape at a constant speed past the magnetic heads while recording or playing back. (Small speed variations, called flutter and wow, do exist, as is discussed later.) All tape transports have provisions for fast winding of the tape, either for reaching a certain portion of the tape in a short time or for complete rewinding.

The tape transport mechanism in its simplest and most reliable form consists of a metal plate with three motors mounted on it, one motor for driving the tape at a constant speed and the other two for reeling the tape. Savings in the manufacturing costs of a tape transport are sometimes sought by using only one motor with friction or clutch drives for the reeling mechanism. The layout in general is as shown in Fig. 2-1, where the tape passes from the supply reel through the guide and over the heads; the tape speed is controlled by the capstan, against which the tape is held by a rubber pinch roller. The capstan shaft rotates at a constant speed and is a critical item in the transport mechanism. It must rotate in good bearings and should ideally be perfectly concentric, with a surface of controlled smoothness. (A perfectly smooth capstan surface will cause tape slippage and associated speed variations.)

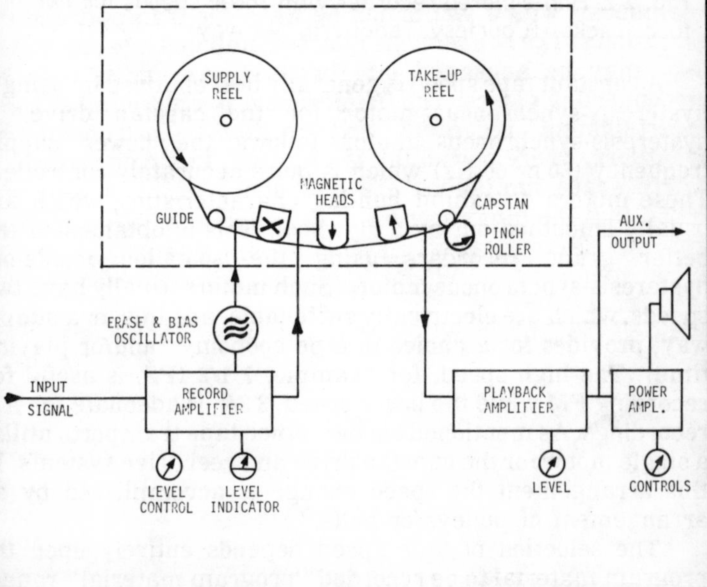

Fig. 2-1. Block diagram depicting the basic elements in a typical tape recorder.

Fig. 2-2. Home stereo recorder with three speeds for two or four tracks. (Courtesy Tandberg, Norway).

A constant tape speed is generally best obtained by using a hysteresis-synchronous motor for the capstan drive. A hysteresis-synchronous motor follows the power supply frequency (50 or 60 Hz) which is quite accurately controlled. These motors do exhibit hunting characteristics, which are usually smoothed out by a flywheel. This is obtained in the better grade recorders using the so-called inside-out hysteresis-synchronous motors. Such motors usually have two speeds, which are electrically switchable, and this, in a simple way, provides for a choice in tape economy (and/or playing time). The high speed, for example, 7 1/2 IPS, is useful for recording FM while the lower speed, 3 3/4, is adequate for AM recordings. As mentioned earlier, other tape transports utilize a single motor for the capstan drive and reel drive systems. In this arrangement the speed change is accomplished by an arrangement of pulleys or belts.

The selection of tape speed depends entirely upon the program material to be recorded "program material" ranges from direct current to a few Hz for underwater oceanographic investigations; two-to-three thousand Hz for taped letters and voice communication; to 15,000 Hz for high-fidelity music reproduction; 2 MHz for wideband recordings; and up to 10 MHz for color video recording.

Since most users of magnetic tape recorders have requirements for one or the other types or ranges of recording, it is generally found that recorders have from two up to six speed ranges. In home entertainment recorders, this is achieved by a two- or three-speed hysteresis-synchronous motor or by an appropriate pulley arrangement, and in instrumentation recorders by the use of servo-controlled tape drive systems.

The tape transport mechanism, in addition to providing a constant speed, must insure good contact between the magnetic tape and the heads. This contact requires a certain amount of pressure, which can be obtained either through hold-back tension on the supply reel, a felt pad against the ingoing tape guide or felt pads which press the tape against the heads.

The felt pad arrangement is inexpensive but can easily cause excessive wear of the heads, which increases the cost of replacing worn-out heads. The felt pad is best used against the tape at a guide post and will, if properly adjusted, give a more constant speed throughout a reel of tape than the hold-back

Fig. 2-3. Home/office two-track recorder. (Courtesy Uher, W. Germany).

torque applied at the feed reel. In a single-motor tape deck, torque is provided by a slip-clutch arrangement and in a three-motor deck by reverse current to the feed-reel motor.

WOW AND FLUTTER

No capstan drive system is perfect and as mentioned earlier, this will give rise to tape speed variations that affect the playback quality of music. (Its effect upon instrumentation data is discussed in Chapter 5.) Even minute speed variations are noticeable, although it does depend on the type of program material. The human ear is particularly sensitive to speed variations during playback of pure tones like bells, flute, and piano. Such speed changes are called wow and flutter. Speed variations up to 10 Hz are called "wow," and speed variations above 10 Hz are considered "flutter."

Wow is primarily caused by capstan shaft eccentricity or dirt buildup on the capstan, or possibly a poor reel drive system with a varying hold-back. Motor cogging and layer-to-layer adhesion in the tape pack can cause wow, too. In cases of wow, tones are frequency modulated which gives rise to a singing sound. Flutter, on the other hand, is generated by the tape itself. Magnetic tape has elastic properties and when it moves over guides and heads, a scraping, jerky motion takes place and this causes longitudinal oscillations in the tape. This destroys the otherwise pure tones and gives them a raw and harsh sound which is often mistaken for modulation noise.

Typical amounts of wow and flutter (up to 250 Hz) that can be tolerated are:

Speech	Max. 0.6 percent
Popular music	Max. 0.3 percent
Classical music	Max. 0.15 percent
Wow is just barely noticeable at	0.1 percent

TECHNICAL SPECIFICATIONS

The wide range of magnetic recorders very often makes it difficult to select the proper unit. First, it is necessary, of course, to establish what the recorder will be used for and in what environment. Magnetic recorders range from miniature, portable, and battery-driven to large instrumentation types.

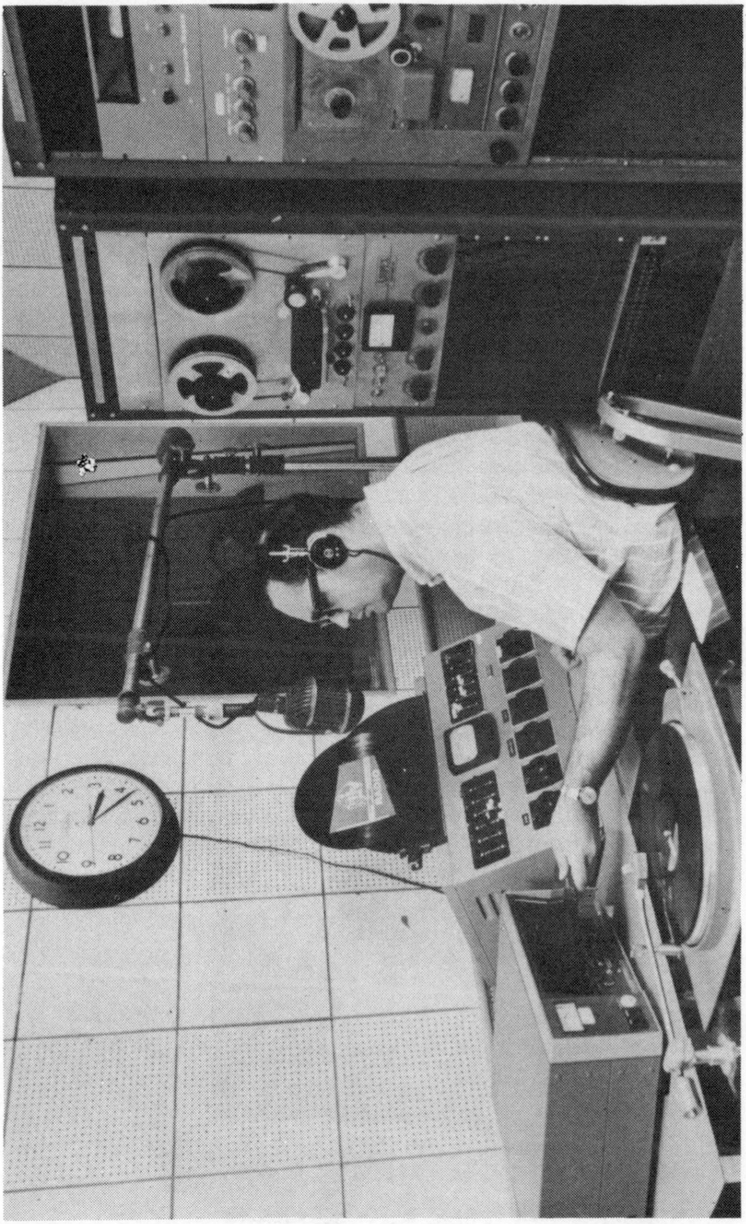

Fig. 2-4. Professional studio recording equipment. Left: Record-playback unit for magnetic discs (spot announcements, short records, etc.) Right: An Ampex 350 and 601-2 recorder. (Courtesy Ampex Corp., U.S.A.)

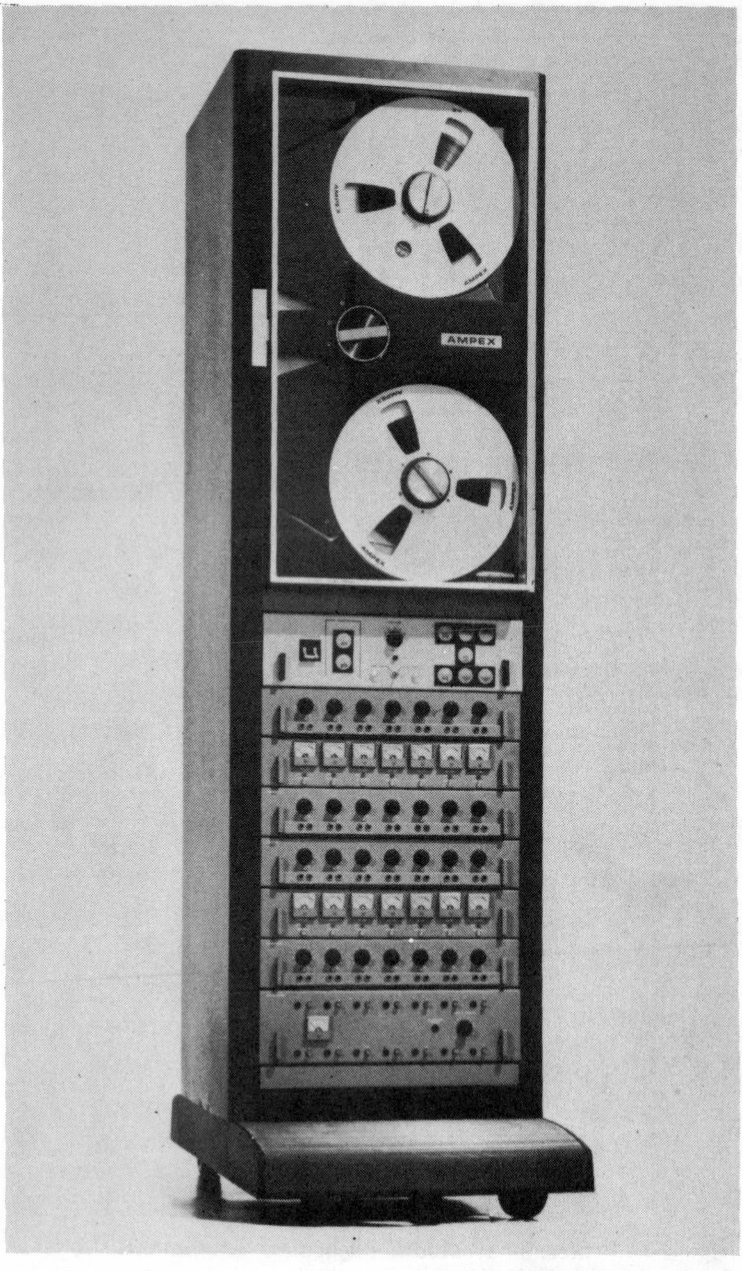

Fig. 2-5. 14-channel instrumentation recorder. (Courtesy Ampex Corp., U.S.A.)

This book is not intended for and cannot cover all the various types and thus is limited to illustratations of only four different types of recorders which are shown in this Chapter. The potential buyer of a recorder should investigate the market by visiting component dealers and by studying data sheets and trade magazines, which for the hobbyist are the high-fidelity magazines.

When such individual choices as price, appearance, and features of the recorder have been satisfied, attention should be given to the technical specifications. In Chapter 1 and in the

Fig. 2-6. 7-channel instrumentation recorder. (Courtesy Hewlett-Packard, U.S.A.)

Fig. 2-7. Digital tape transport for computer systems. (Courtesy Hewlett-Packard, U.S.A.)

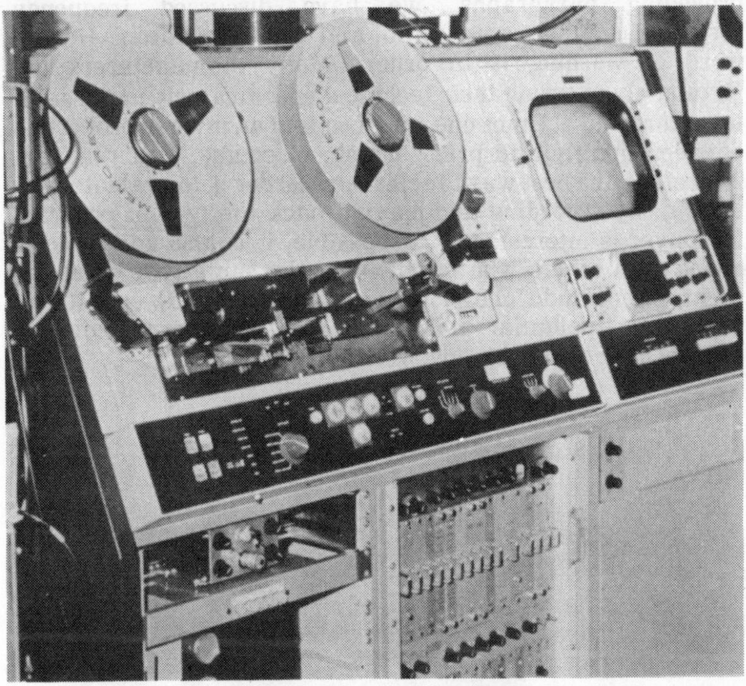

Fig. 2-8. Television recorder. (Courtesy Ampex Corp., U.S.A.)

Fig. 2-9. Compact, high-altitude environment instrumentation recorder. (Courtesy Astro-Science Corp., U.S.A.)

preceding paragraphs, we have discussed frequency response, signal-to-noise ratio, and wow and flutter. Here a word of warning is in order: Many manufacturers unfortunately overrate their technical specifications in the sales literature and at times one can even find an inverse proportion between quality and price, which, of course, is unrealistic. Therefore, the best way to select a recorder is to evaluate it by testing; by recording and playing back the type of program the buyer is interested in. If possible, it is also advisable to check with others who might have the same type of equipment. To conclude, one should be advised as to how easy it is to repair any potential malfunction in the recorder and, of course, he should know how well it is constructed.

Chapter 3

The Transport

A wide range of tape-handling mechanisms are in existence today, ranging from the simple, cartridge players used in automobiles, home recorders, professional recorders, and sophisticated instrumentation recorders. These various tape transport mechanisms are essentially divided into two "families": one for analog recording, such as music or data information where the tape is moved at a constant speed; the other "family" includes the computer digital transport where the tape is moved at a high speed and is stopped in rapid sequences. Since the computer digital transport is basically an outgrowth of the analog transport, the emphasis here is placed on the analog tape drive. Considerations for proper tape handling apply equally to both types, which in essence means that neither type of transport should cause any physical damage to the magnetic tape.

LAYOUT

The functional layout of a magnetic tape transport is dictated by its application. Fig. 3-1 shows four different layouts used for analog transports. Fig. 3-1A illustrates the reel-to-reel concept which is so commonly found in home-type recorders and in many instrumentation-type recorders. Fig. 3-1B shows the compact version where one reel is located concentrically above the other, a feature which saves space. A cartridge design–Fig. 3-1C–such as that used in many automobile installations is also shown, together with an endless loop contained in a bin (Fig. 3-1D). The latter concept is often used where a constant monitoring of data is required without the need for prolonged storage, which is exemplified in flight recorders where only the last few minutes of information must be stored.

In a digital computer transport the tape motion must start and stop almost instantaneously, and this is achieved by a

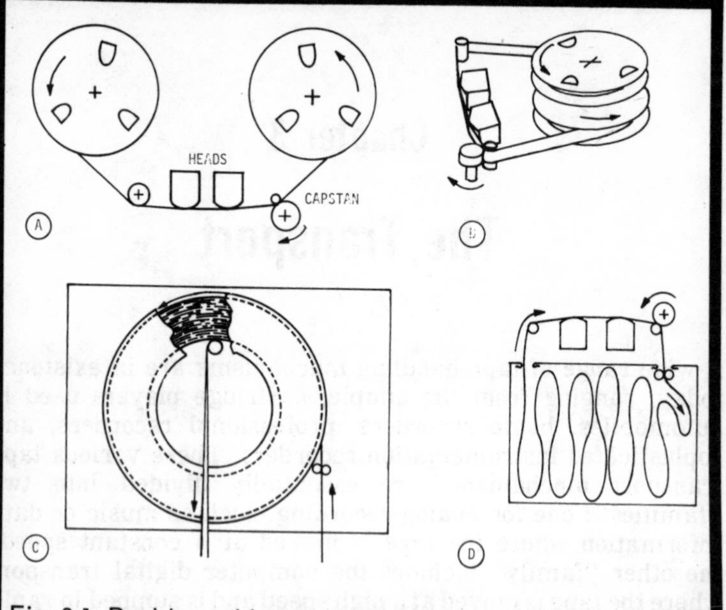

Fig. 3-1. Reel-to-reel transport, side by side (A), and stacked (B). Endless loop cartridge (C), and bin-storage type (D).

layout similar to that shown in Fig. 3-2 where two long tape loops—one from each reel—are stored in vacuum-controlled bins. This means that the capstan drive, which is the speed-controlling element in a tape transport, has only to overcome the inertia of the free-hanging tape loops rather than that of a tape reel.

For video or television recordings, two concepts are used. Fig. 3-3A shows the broadcast version, which is the original Ampex concept, using a 2" wide tape with "transverse scanning," where recording and playback are accomplished by a rotating-head assembly. A more recent version of television recorders, aimed particularly at school use and potentially for home use, is shown in Fig. 3-3B, which utilizes what is called the helical-scan technique. Here, again, the magnetic-head assembly rotates but the recorded pattern on the tape is stored at an angle.

DRIVE SYSTEMS

The most important element in a magnetic tape transport is the capstan—a precision drive shaft against which the tape is

Fig. 3-2. Most computer transports use vacuum columns to house tape loops.

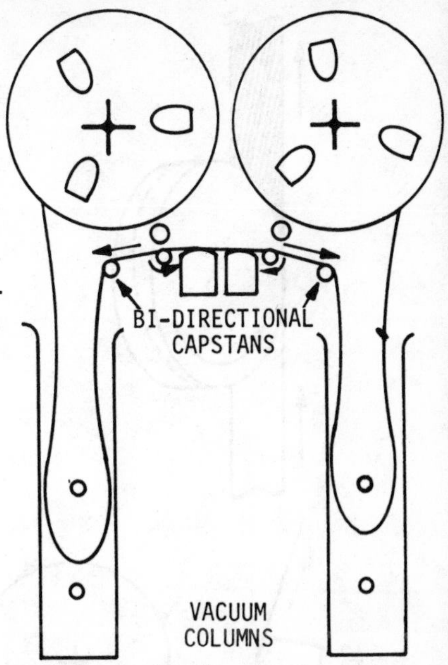

pressed by a pinch roller. The speed of the capstan determines the tape speed. The diameter of the capstan shaft varies from about 1/8 of an inch for slow-speed home recorders to about 1 inch for high-speed, precision instrumentation recorders.

There are essentially three basic capstan drive systems, as shown in Fig. 3-4. The oldest and still a much used concept (Fig. 3-4A) is the open-loop drive where the capstan pulls the tape over the head with tension provided by hold-back torque or a felt-pad arrangement. These transports often have an idler which provides isolation from any tape projections from the supply reel. As regards isolation, the closed-loop drive, shown in Fig. 3-4B, is superior in isolating any speed variations from the reeling system and this concept is found in several instrumentation recorders. A more recent and superior drive system is shown in Fig. 3-4C, where two capstans with associated pinch rollers are used. One capstan rotates at a slightly higher speed than the other, thereby providing tension on the heads. Both capstans could rotate at the same speed with the diameter of one capstan being slightly larger than the other.

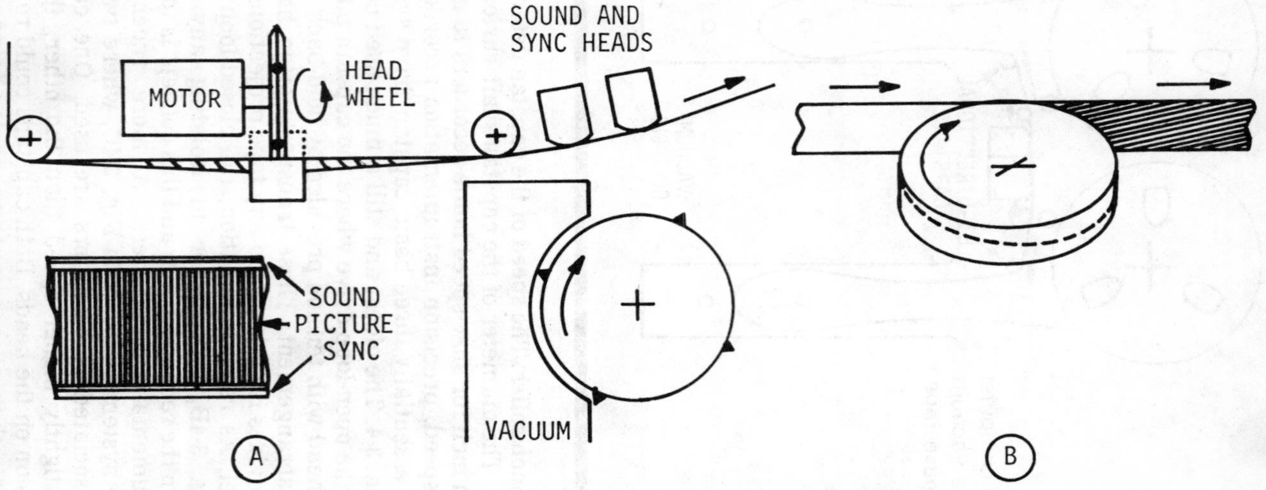

Fig. 3-3. Video transports. A. Transverse scan for 2″ tape. B. Helical scan for 2″, 1″, and ½″ tape.

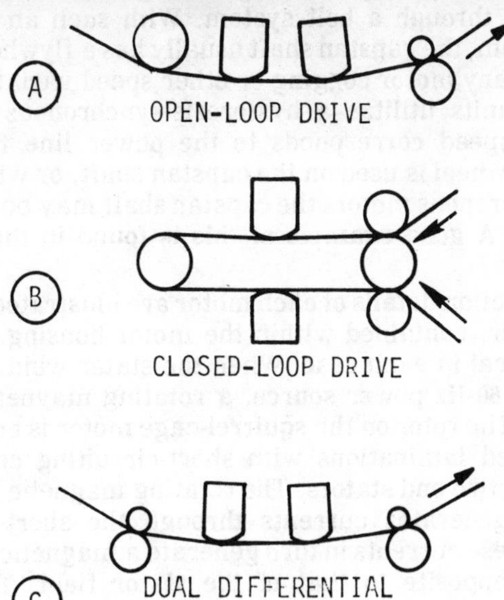

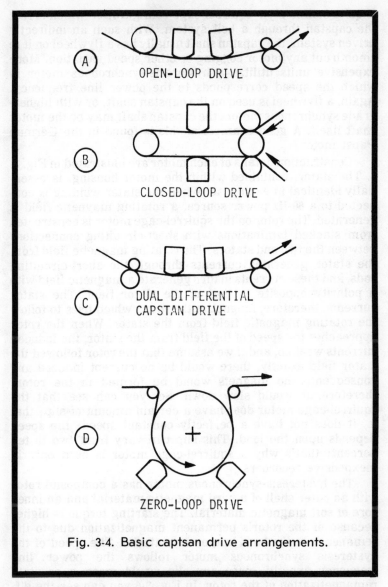

Fig. 3-4. Basic captsan drive arrangements.

Fig. 3-4D illustrates another drive concept, where the tape contact with the large diameter capstan is held by a vacuum. This concept is referred to as a zero-loop drive and has the advantage that the tape is supported all the way past the heads, which eliminates scrape flutter.

The capstan shaft is driven by either an AC or a DC motor. In inexpensive home entertainment recorders the drive motor is quite commonly a squirrel-cage configuration which drives the capstan through a belt system. With such an indirectly driven system, the capstan shaft usually has a flywheel on it to smooth out any motor cogging or other speed variation. More expensive units utilitze a hysteresis-synchronous motor in which the speed corresponds to the power line frequency. Again, a flywheel is used on the capstan shaft, or with higher-grade synchronous motors the capstan shaft may be the motor shaft itself. A good example of this is found in the German Papst motor.

Construction details of each motor are illustrated in Fig. 3-5. The stator, contained within the motor housing, is essentially identical in each case. When the stator winding is connected to a 60-Hz power source, a rotating magnetic field is generated. The rotor on the squirrel-cage motor is constructed from stacked laminations with short-circuiting connections between the two end stators. The rotating magnetic field from the stator generates currents through the short-circuiting rods, and these currents in turn generate a magnetic field with a polarity opposite to that of the stator field. The stator currents, therefore, magnetize the rotor, which tries to follow the rotating magnetic field from the stator. When the rotor approaches the speed of the field from the stator, the induced currents weaken, and if we assume that the rotor followed the stator field exactly, there would be no current induced and consequently no magnets would be formed in the rotor; therefore, it would slow down. So, you can see that the squirrel-cage motor does have a certain amount of slip; that is, it does not have a perfectly constant speed; the speed depends upon the load. This slip can vary from two to ten percent; that's why a squirrel-cage motor is used only in inexpensive recorders.

The hysteresis-synchronous motor has a composite rotor with an outer shell of a hard magnetic material and an inner core of soft magnetic material. The starting torque is higher because of the rotor's permanent magnetization due to the remanence in the hard magnetic material. The speed of the hysteresis synchronous motor follows the power line frequency exactly, since any slip would mean a constant remagnetization of the rotor. In Fig. 3-6 you can see the difference in the speed vs torque curves for the two types of motor. It is evident that the hysteresis-synchronous motor is superior, both with regard to constant speed and to a higher starting torque.

As mentioned above, most recorders are provided with two and sometimes three speed ranges. This is achieved in two

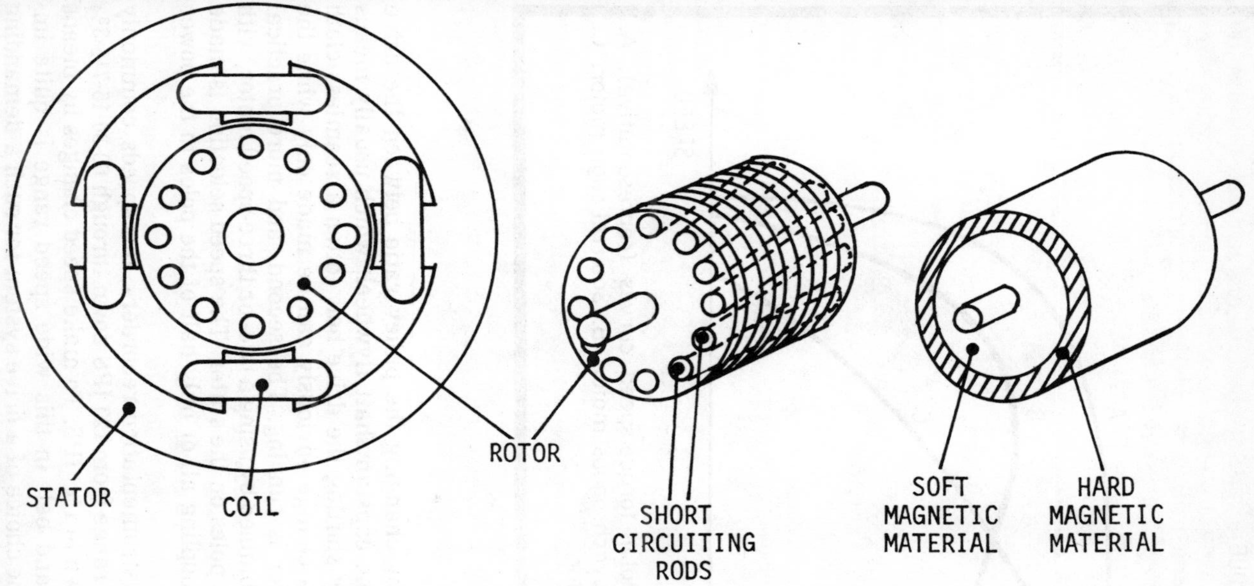

Fig. 3-5. Construction details of a squirrel-cage motor. The right-hand drawing shows the rotor of a hysteresis-synchronous motor armature.

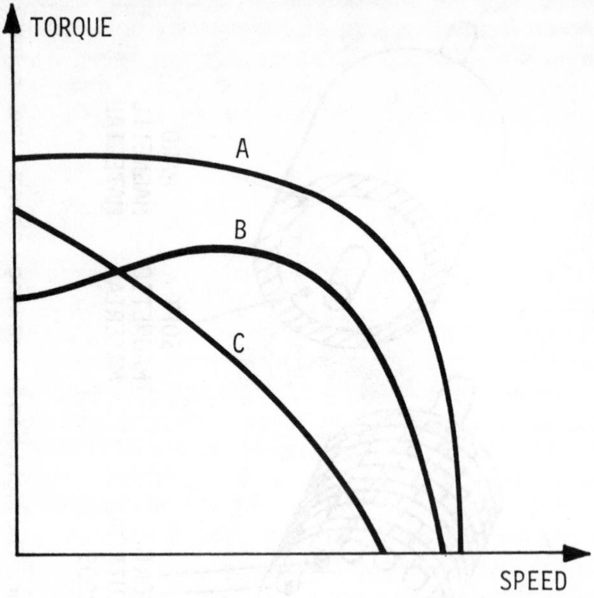

Fig. 3-6. Motor torque-speed curves (representative). A. Hysteresis-synchronous motor. B. Squirrel-cage motor. C. Torque motor.

ways: one, by changing the pulley ratio between the drive motor and the capstan shaft flywheel, which usually means mechanically shifting the drive belt (often a seamless cloth belt). Such a change obviously can be made only while the capstan motor is running. The second, and more practical solution, is obtained by using a two- or three-speed motor, with four or eight poles on the stator. The speed selection is made by merely coupling all or only half of the poles to the power source.

In the instrumentation recorders, the speeds commonly encountered range from 120 IPS down through 60-30-15-7½-3¾ and even down to 1⅞ IPS. To make speed changes by means of a pulley and belt in this wide speed range is quite impractical. The choice of a drive system for such a demanding system, therefore, falls on a servo-controlled system, where the capstan drive motor is a DC type with a suitable tachometer mounted on the drive-motor shaft. The output frequency of the tachometer is compared with a control reference frequency, and any frequency deviation results in an error signal which corrects the motor speed. Here, again, the drive motor can be coupled to the capstan shaft by means of a belt drive or the capstan shaft can be the motor shaft itself.

FLUTTER AND TIME-BASE ERROR

In earlier instrumentation recorders, speed control is attained by recording a 17-kHz controlled frequency which is amplitude-modulated at a rate of 60 Hz. During playback, this signal is demodulated and fed into a phase-comparison circuit, which feeds a precision 60-Hz motor-control oscillator whose output after amplification drives the synchronous capstan motor. This technique is referred to as **speed control**, since the servo system functions through a high-inertia drive system with flywheels and belt drive and the maximum error rate that can be followed is around 1 Hz.

Instrumentation recordings are often made under adverse conditions (vibrations, etc.) and the inherent wow and flutter both from the recorder and the external vibrations in turn affect the frequency of the recorded tones. As will be shown a little further on in the chapter, this will have an effect on any analysis work that may be performed from the data stored on the tape. The change of speed causes a change of tone frequency, which is translated to a change in timing. Therefore, it is vital that this error, also called **time-base error**, be reduced. This requirement has led to a new family of instrumentation recorders which have a drive system as illustrated in Fig. 3-7. Here, the tape-drive motor is a low inertia, high-peak-torque motor, driven by a power amplifier controlled by a servo amplifier. When the tapes were recorded originally, one track was recorded with a control frequency ranging from 200 kHz down to 6.25 kHz, all dependent upon the speed selected. Upon playback this signal is amplified and fed to a phase detector and compared with a crystal-controlled control frequency. Any speed deviation results in a phase error signal which in turn controls the servo amplifier and causes the tape drive to speed up or slow down. The inertia of the capstan motor, and possibly the coupling elements, restrict the time for making speed corrections; a usual figure of merit is **saturation acceleration** (equal to the peak torque of the drive motor divided by the inertia). A low inertia drive with a high saturation acceleration is far more capable of

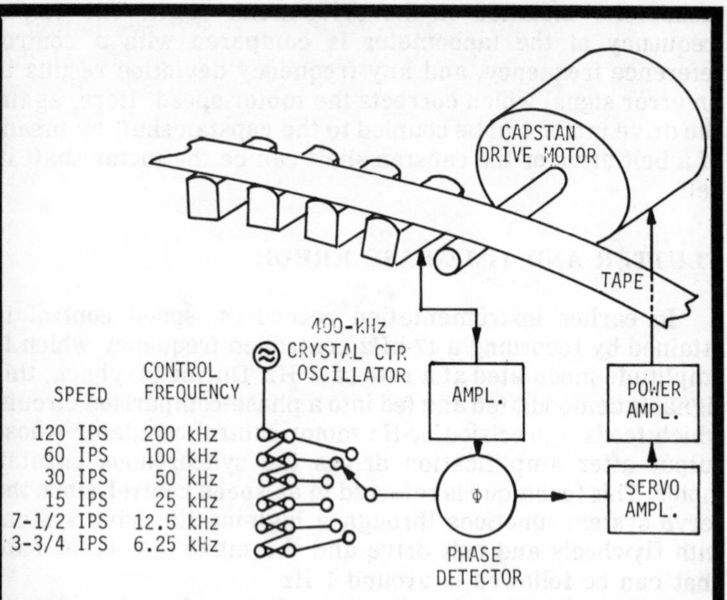

Fig. 3-7. Pictorial diagram of a typical servo speed control system.

correcting speed variations than any of the more conventional transports with their heavier motors and flywheels.

We can calculate a typical example for the required acceleration in a case where a tape has been recorded at 60 IPS with a disturbing flutter frequency of 15 Hz with a magnitude of 1 percent peak-to-peak. This can be expressed mathematically:

$$V = (60 + .3 \sin\omega t) \text{ IPS}$$

By differentiating the speed, we find the acceleration:

$$\frac{dv}{dt} = .3\omega \cos\omega t \text{ inches/sec}^2$$

$$\frac{dv}{dt\text{max}} = .3 \times 2\pi \times 15$$

$$= 28 \text{ inches/sec}^2$$

This is far in excess of what a conventional recorder with speed control will handle—it corresponds to a start time of 1 second for 30 IPS operation with no overshoot. Since flutter is never purely sinusoidal, the maximum acceleration for the example may well be on the order of several hundred inches/sec^2, which the reproducer must follow so as not to lose synchronization.

The specification most often encountered for these new recorders is TBE equals Time-Base Error (also time displacement error), which gives the residual dynamic time difference from "real" time. In a low TBE recorder/reproducer the timing error is reduced by an order of two magnitudes, but the flutter is not reduced by the same amount. This is illustrated in Fig. 3-8, where a typical flutter

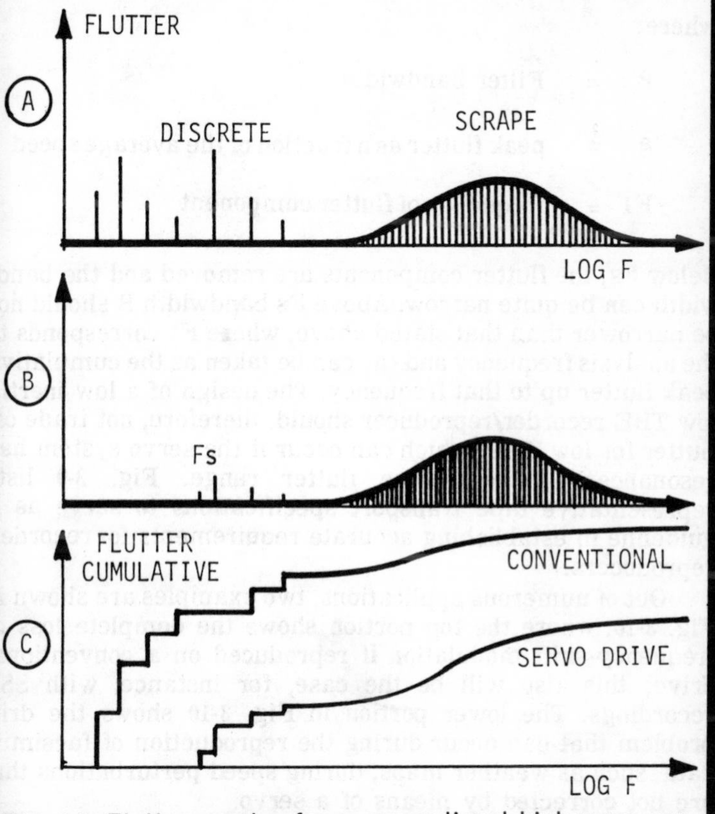

Fig. 3-8. Flutter spectra for a conventional high-mass capstan drive (A) and a low-mass servo-controlled drive (B). The bottom curves (C) show the cumulative results for A and B.

spectrum is shown for a conventional transport, consisting of a number of discrete components and a random distributed scrape flutter. When a servo-mechanism is employed, it removes the components below Fs, which is the upper frequency response of the electro-mechanical servo, and when the cumulative flutter is measured (DC to 10 kHz) there is a reduction of flutter, but this does not correspond to the reduction in TBE, except below Fs.

In spectrum analysis work, it has been shown that the criteria[3] for full recovery of the power in the signal is:

$$B \geqq 3 (aF1)$$

where:

B = Filter bandwidth

a = peak flutter as a fraction of the average speed

F1 = frequency of flutter component

Below Fs, the flutter components are removed and the bandwidth can be quite narrow. Above Fs bandwidth B should not be narrower than that stated above, where F1 corresponds to the analysis frequency and (a) can be taken as the cumulative peak flutter up to that frequency. The design of a low inertia low TBE recorder/reproducer should, therefore, not trade off flutter for low TBE, which can occur if the servo system has resonances in the scrape flutter range. Fig. 3-9 lists representative tape transport specifications to serve as a guideline in establishing accurate requirements for recorder/reproducers.

Out of numerous applications, two examples are shown in Fig. 3-10, where the top portion shows the complete loss of frequency-shift modulation if reproduced on a conventional drive; this also will be the case, for instance, with SSB recordings. The lower portion in Fig. 3-10 shows the drift problem that can occur during the reproduction of facsimile data, such as weather maps, during speed perturbations that are not corrected by means of a servo.

REELING SYSTEM AND TAPE GUIDANCE

While the capstan drive system plays a major role in maintaining speed accuracy, emphasis should be placed upon

INERTIA DRIVE	FLUTTER	TBE	SATURATION ACCELERATION	Fs-SERVO CUT-OFF FREQ.
HIGH-INERTIA DRIVE W/SPEED CONTROL	.6 % P-P	±100μs	$1 \frac{INCHES}{SEC^2}$	1 Hz
MEDIUM-INERTIA DRIVE W/SERVO CONTROL	.45% P-P	± 1μs	$100 \frac{INCHES}{SEC^2}$	100 Hz
LOW-INERTIA DRIVE W/SERVO CONTROL	.25% P-P	± .5μs	$1000 \frac{INCHES}{SEC^2}$	1000 Hz

Fig. 3-9. Representative transport specifications.

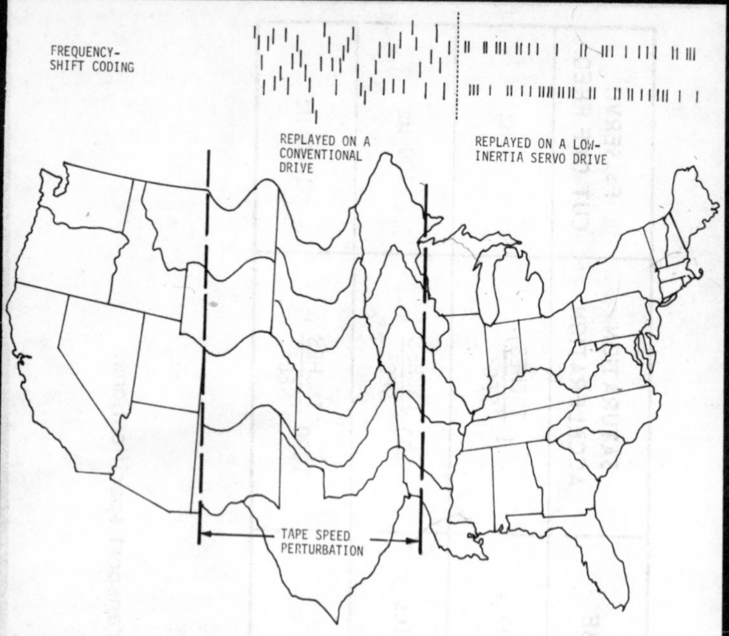

Fig. 3-10. Effects of time-base error on a coded signal and a facsimile transmission.

the reeling system also. The **pay-off reel** should supply the tape to the head area in a perfectly smooth motion, while the **takeup reel** should wind the tape onto the reel under constant tension and with a smooth tape pack.

During winding or re-winding, when the capstan puck is disengaged, the tape should move with the highest practical speed from one reel to the other, and when the winding or re-winding mode ends, both reels should stop in such a fashion that the tape will neither be overstretched nor throw a loop. This again puts a demand on the design of the braking mechanism for the reel system.

The best reeling is obtained by the use of two separate reel motors, which should be **torque motors** that provide for an almost constant tape tension. Such motors have an inverse speed-torque curve, as shown in Fig. 3-6. When the reel is empty the speed is high, the torque is low and a given tape tension is provided. On the other hand, when the reel is filling up, the speed slows down and the torque increases, but the radius of tape accumulating on the takeup reel is increasing and the tension is essentially constant from the empty to the full reel. The almost constant tension is desirable because when magnetic tapes are stored for a long period under excessive tension or a highly varying tension through the reel, the result can be a permanent distortion of whatever is recorded on the tape.

Constant tension is also important for unvarying tape speed, in particular with open-loop transport drives. If the tension throughout a reel varies, it will affect the speed and, in poorly designed recorders, the absolute speed of the tape can change several percent from the beginning to the end of a recording. If played back on the same recorder, unequal tension will not lead to any change in quality. But if the tape is edited, and especially if a portion is moved from the beginning to the end or vice versa, there will be a noticeable change in the frequency or pitch of the recorded material.

The braking system, which (as earlier mentioned) works in conjunction with the reeling motors, must be carefully designed to avoid tape damage. An often used braking system is shown in Fig. 3-11. Each brake consists of a flat steel band with a felt coating. The braking action is differential; that is, if the tape is traveling from left to right when the brakes are engaged, the left wheel will receive a slightly greater braking action than the right wheel, and vice versa. This braking system is disengaged during playing or winding either by means of a lever or by solenoids. The use of solenoids is a more

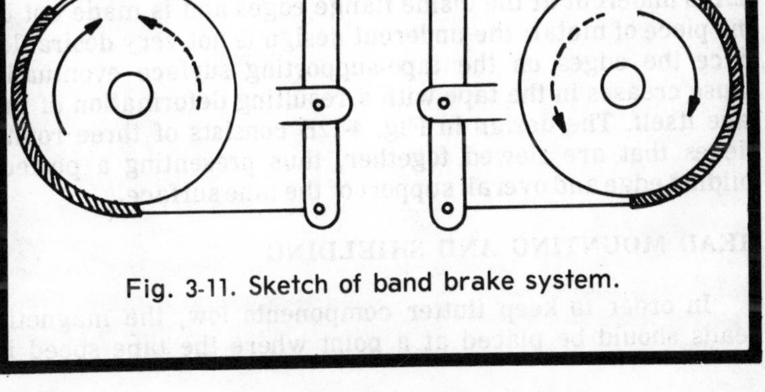

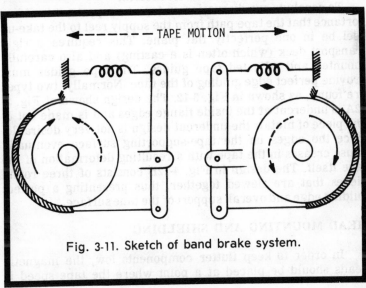

Fig. 3-11. Sketch of band brake system.

attractive solution since any power failure to the recorder will cause the brakes to automatically engage.

Spring loading for the brakes must also be given consideration during the design of a recorder. Most home recorders are designed for the use of 7-inch reels, but quite often such recorders are used with, say, a 7-inch reel for takeup and a 3-inch reel for supply (in the exchange of tape letters, for example). Here, the dynamics of the braking action are different from that originally designed for two reels of the same size and the operator should pay great care in starting and stopping a recorder with different reel sizes. The same problem exists in instrumentation recorders where the two commonly used reel sizes are 10½- and 14-inch diameter reels. The larger reel has far more inertia than the smaller reel; therefore, some of these recorders are equipped with a switch that selects the proper braking action for the various reel sizes.

Flutter is quite often introduced into a recording by either eccentric reels, an improper hold-down mechanism for the reels, or by cogging in the reel motors. These flutter components are particularly noticeable when the reel is almost empty and they can be avoided only by a careful design of the reel hold-down mechanism. Several instrumentation manufacturers incorporate sensing arms in the tape path; the arms have transducers that detect any variations, thereby compensating for any reel rate or reel motor cogging by servo-controlling the speed of the reel motors.

To conclude our discussion on tape transports, it is of importance that the tape path from the supply reel to the take-up reel be in one perfectly flat plane. This requires a rigid transport deck (which often is a casting) and also carefully mounted and adjusted tape guides. The tape guides must provide perfect edge guiding of the tape. Normally, two types are found, as shown in Fig. 3-12. The design shown at Fig. 3-12A is undercut at the inside flange edges and is made out of one piece of metal; the undercut design is not very desirable, since the edges on the tape-supporting surface eventually cause creases in the tape with a resulting deformation of the tape itself. The design in Fig. 4-12B consists of three round pieces that are slewed together, thus presenting a perfect guiding edge and overall support of the tape surface.

HEAD MOUNTING AND SHIELDING

In order to keep flutter components low, the magnetic heads should be placed at a point where the tape speed is smoothest and this is normally as close to the capstan as possible. Attention should be paid here to the fact that the magnetic tape itself is an elastic medium and that longitudinal oscillations (similar to those of a violin string) occur. Whenever the tape runs over a fixed guide or magnetic head, it rubs against these fixed elements, resulting in longitudinal oscillations (scrape flutter). Guide and head surfaces must be as perfectly smooth as possible, to minimize flutter, and the wrap angle around the heads should be a minimum, normally in the order of 3 to 5 degrees. If the heads are located between the idler and the capstan, scrape flutter will be reduced. A rotating high-inertia guide will likewise serve as a grounding point for the longitudinal oscillations, and such guides are frequently incorporated into precision tape drives.

In all cases of longitudinal tape oscillations there is a natural resonance frequency, which by a rule of thumb is 40,000 Hz at one inch. So, if we have a free tape span from the in-going idler to the capstan of say 5 inches, the scrape flutter resonant frequency will be most noticeable at 8,000 Hz. If, on the other hand, the tape span is 10 inches, the resonant frequency will be 4,000 Hz. Thus, a short tape span is important. In the design of a precision tape recorder, it is quite useful to use the equivalent electrical diagram for the electromechanical components of the recorder. (An excellent example of this equivalence has been published; see reference 4 at the end of the Chapter.)

HEAD MOUNTING AND SHIELDING

In magnetic recorders with low packing densities, it is quite common to find that the heads are mounted on a fixed base plate, which in turn is mounted on the capstan drive precision plate. Where the packing densities (i.e., wavelengths per linear inch of tape) approach 20,000 Hz per linear inch of tape, it becomes necessary to mount the reproduce head on an alignment plate so its azimuth can be tilted to be exactly equal to that of the record head. The record heads are normally fix-mounted with a certain tolerance given for the azimuth which is obtained by careful grinding of the recording head mounting surface in the final step of its production.

Reproduce heads must be carefully shielded with a mu-metal housing or another suitable metal in order to minimize hum pickup from the drive motors, transformers, and any other external interference source. In wideband recorders

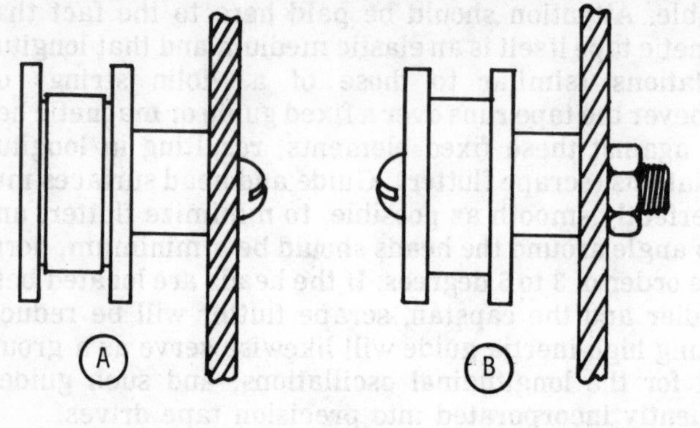

Fig. 3-12. Tape guides constructed as shown at A tend to deform the tape.

that operate in the range of up to 1.5 MHz, additional shielding becomes necessary in order to avoid pickup of signals in the midwave radio-frequency band. Powerful broadcast stations in this range may be picked up by a sensitive reproduce head and appear as noise in the recorder output. In the design of the head mounting, attention should also be paid to the fact that signals from the record head may radiate into the reproduce head during recording, which makes alignment of the recorder's electronics difficult.

References:

1. Athey, Skipwith W.: "Magnetic Tape Recording," NASA SP-5038. For sale by the Superintendent of Documents, U.S. Government Printing Office, Washington D.C., 20402, U.S.A.

2. James, E.R.: "Magnetic Tape Transports for System Build-In Applications," Electronic Instrument Digest, April 1967, p. 40-48.

3. Jorgensen, Finn: "The Accuracy of Analog Magnetic Recording Storage," Telemetry Journal, June/July 1966, p. 41-44.

4. Wolf, W.: "An Investigation of Speed Variations in a Magnetic Tape Recorder With the Aid of Electro-Mechanical Analogies," Audio Engr. Soc., Vol. 11, April 1960, pp. 119-129.

Chapter 4

Magnetic Heads & Tapes

Elementary magnetic recording and playback principles were introduced in Chapter 1, where some practical aspects were highlighted. Now, let's explore the physics behind recording and playback so as to arrive at the requirements for magnetic heads and tapes. Fig. 4-1 is a block diagram of the head transducers and electronics in a typical magnetic recorder/reproducer. In Fig. 4-1 emphasis is placed upon the magnetic tape and the heads.

The **erase head** is supplied with a strong alternating current from an erase oscillator. The oscillating magnetic field generated by the head saturates the tape, and as it leaves the erase head the field decays and the tape is demagnetized. The erase head in Fig. 4-1 shows a double front gap; this arrangement does a better erasing job than a single gap. The gap in the back of the core is large to prevent the core from becoming permanently magnetized. A narrow back gap would act as a keeper for any permanent magnet or "perm" as it is called in the industry.

When the tape moves past the **record head** it passes through a field that magnetizes the coating in accordance with input signal Vin. The recording field is a superimposition of two fields: a data field and a bias field. The data field is generated by amplifying input voltage Vin to a suitable current, Irecord, which flows through the record-head winding. The bias field, which is approximately ten times stronger than the data field, is supplied from a high-frequency oscillator. The frequency of the bias oscillator is three to five times higher than the highest data frequency that is to be recorded. The bias in audio recorders is commonly supplied from an oscillator that also seves as the current source for the erase head.

The recorded data is played back when the tape passes over the **reproduce head**. The external flux lines from the tape flow through the head core and induce a voltage, Vplay, in the winding. This voltage, after amplification and equalization, represents the recorder input voltage–i.e., Vout is nearly

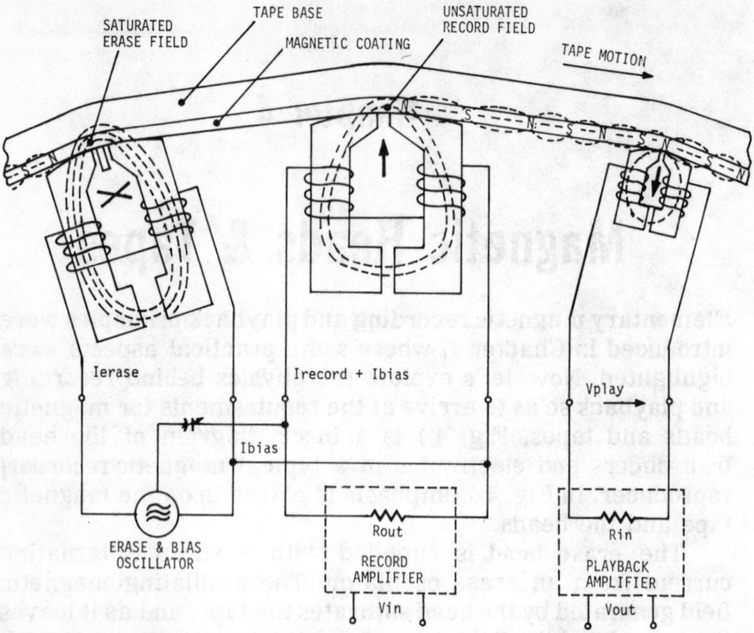

Fig. 4-1. Typical block diagram of the head transducer layout and associated electronics.

equal to Vin, with a time delay due to the spacing between the record and reproduce head.

MAGNETIC RECORDING WITH AC BIAS

Various methods for recording were outlined in Chapter 1. Here, we shall consider only recording with a high-frequency AC bias, as it is used in all high-quality audio and instrumentation recorders. Over the years, numerous attempts have been made to explain the linearizing effect of AC bias. The most accurate thinking is based on the anhysteretic theory (from the physics of magnetization of small particles[1]). Its origin lies in work with fine magnetic particles. When these particles are closely packed, their magnetic fields affect one another and the net effect from this interaction is partly responsible for the action of AC bias.

The hysteresis loop is almost rectangular for a small particle, such as shown in Fig. 4-2A. Such a loop tells that the particle is always magnetized in one direction, plus Br or minus Br. In order to change the magnetization, it is necessary that the field strength H exceeds the coercive force Hc.

When a number of these particles are packed together, as in the tape coating, their fields interact. This is shown in Fig. 4-2B, and each particle is, therefore, influenced by a net field strength we can call delta H. If such a group of particles is subjected to an AC field that slowly decays to zero, the particles will become magnetized in accordance with the interacting fields, delta H, from neighboring particles. This is illustrated in Fig. 4-3, where the particles are subjected to a field that slowly decays. The first particle which is influenced by a negative interacting field minus delta H will after time T1 no longer be remagnetized and will stay at plus Br. The second particle will assume either plus Br or minus Br after time T2; the third particle will no longer remagnetize after time T3 and will stay at minus Br. For example, if we have 300 particles, of which 100 have a negative interaction (minus delta H), another 100 have a positive interaction (positive delta H), and the remaining 100 have zero delta H, then the overall magnetization of the 300 particles will be zero. (On a statistical basis, the 100 particles where delta H equals zero will become 50 with plus Br and 50 with minus Br.)

Present magnetic tapes are manufactured from particles that interfere with each other to a smaller or larger extent, negative or positive. Fig. 4-4A shows how the particle in-

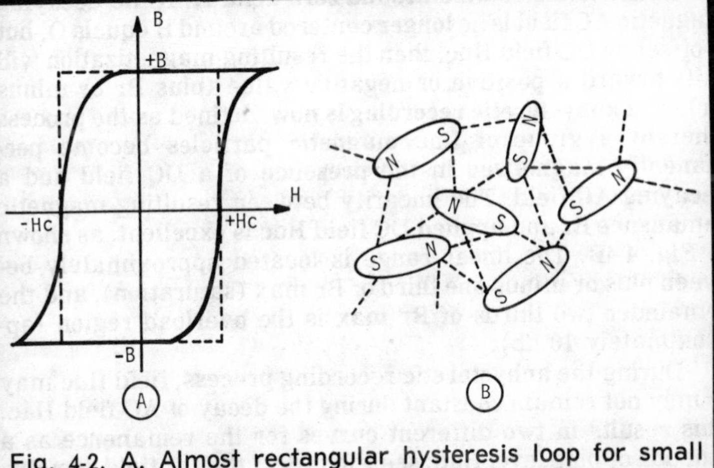

Fig. 4-2. A. Almost rectangular hysteresis loop for small particles. B. Sketch illustrating the interaction between particles.

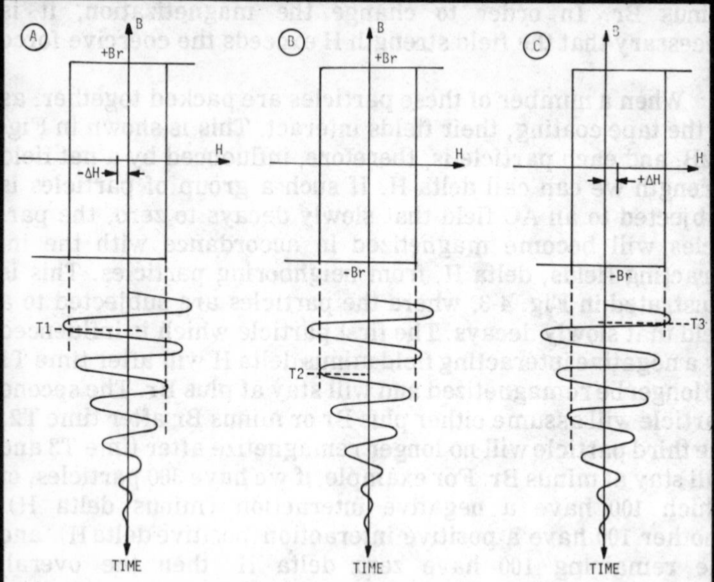

Fig. 4-3. Magnetization of a particle under influence of: (A) negative interacting field -ΔH; (B) no field; (C) positive interacting field +ΔH.

terference is distributed around zero delta H. If the decaying magnetic AC field is no longer centered around H equals O, but is offset by DC field Hdc, then the resulting magnetization will shift toward a positive or negative value (plus Br or minus Br). The **anhysteretic recording** is now defined as the process whereby a group of fine magnetic particles become permanently magnetized in the presence of a DC field and a decaying AC field. The linearity between resulting magnetic remanence Br and applied DC field Hdc is excellent, as shown in Fig. 4-4B. The linear range is located approximately between plus or minus one third of Br max (saturation), and the remainder two thirds of Br max is the overload region (approximately 10 db).

During the anhysteretic recording process, field Hdc may or may not remain constant during the decay of AC field Hac. This results in two different curves for the remanence as a function of bias level Hac; see Fig. 4-5. If the DC field remains constant while the bias field decays, we then have the ideal process where overbias does not cause a decrease in the

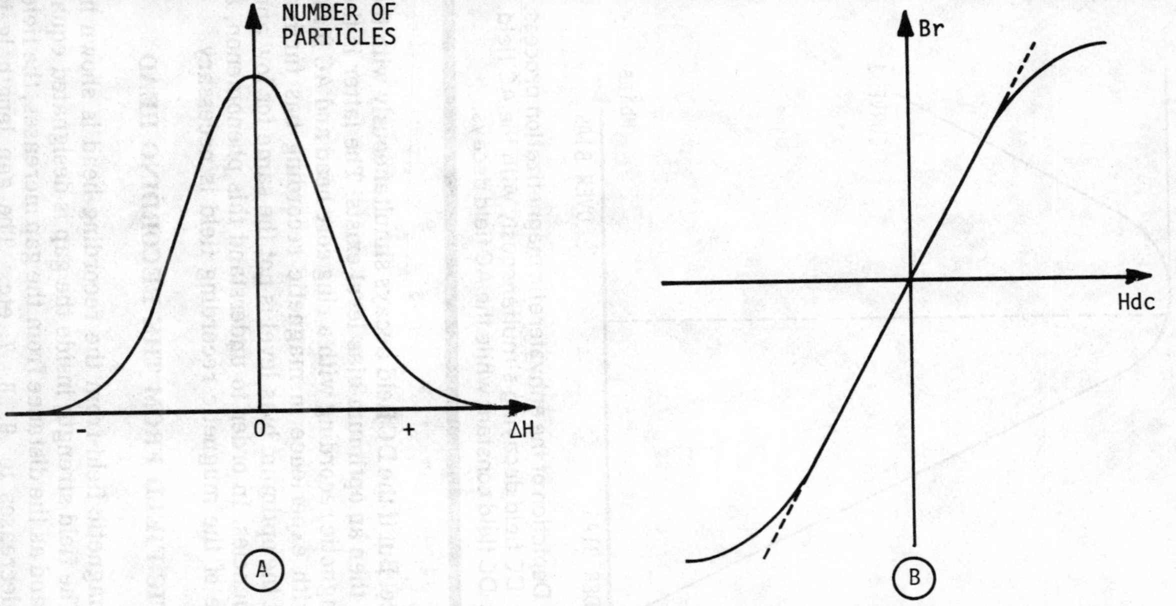

Fig. 4-4. A. Curves showing the distribution of field interaction between particles. B. Remanence Br vs recording field strength in an anhysteretic recording.

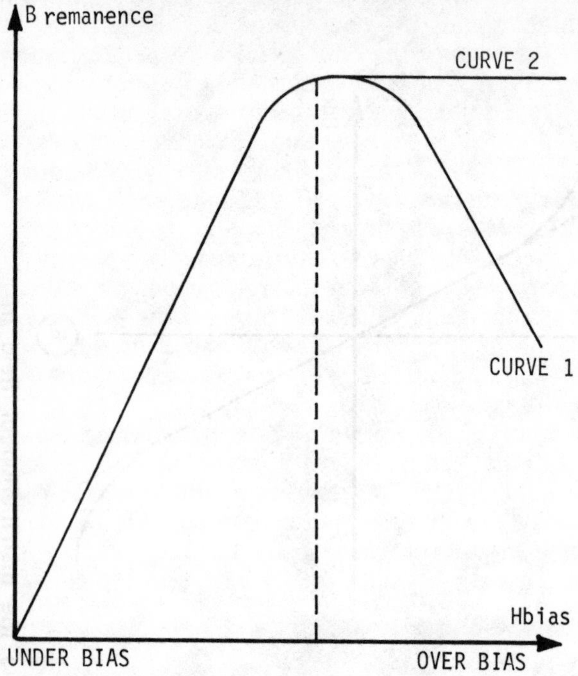

Fig. 4-5. Depiction of the anhysteretic magnetization process. Curve 1: DC field decaying simultaneously with the AC field. Curve 2: DC field constant while the AC field decays.

remanence. But if the DC field decays simultaneously with the bias field, then an optimum bias level exists. The latter is the case in magnetic recording with a ring core head and AC bias. Anyone with experience in magnetic recording has further found that the optimum bias level is not the same for low and high frequencies. In order to understand this phenomenon, a knowledge of the magnetic recording field is necessary.

MAGNETIC FIELD FROM THE RECORDING HEAD

The magnetic field from the recording head is shown in Fig. 4-6. The field strength inside the gap is designated equal to 1 (one) and as the distance from the gap increases, the field strength decreases to .9, .8, .7, etc. The gap length lg is generally in the order of .25 mils (6 mu) and the thickness of the magnetic coating .1 to .4 mils (2.5-10 mu). In order to fully magnetize the magnetic coating, the field strength along the .5 line must equal or exceed a critical field strength which approximately equals the coercivity, Hc, of the tape.

This leads us to the concept of a **recording zone** which, speaking in terms of field lines, has a width of plus or minus delta H around the line corresponding to H equals Hc. This is illustrated in Fig. 4-6 where the recording zone is shown for different bias levels. The top figure shows a recording where the critical zone reaches through the magnetic coating of the tape. When the tape approaches the gap, it enters a magnetic front of increasing field strength, and as it passes over the gap, all particles "flip" back and forth under the influence of a saturating bias field. When the tape passes over the trailing edge of the gap, it goes through the recording zone and becomes permanently magnetized in accordance with the presence of data field Hd.

The width of the recording zone determines the shortest wavelength that can be recorded on the tape. If the polarity of the data field changes 180 degrees as the tape moves through the recording zone, the net recording will be zero. It is necessary, therefore, to have a recording zone as short as possible. This can be achieved by reducing the bias level, as

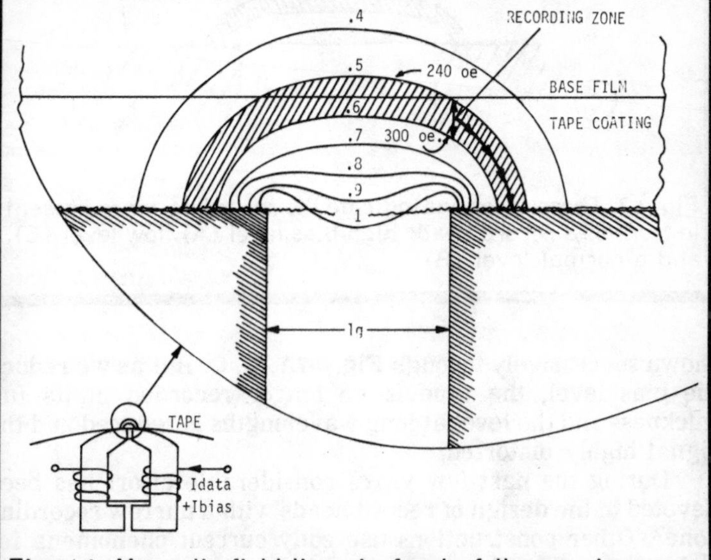

Fig. 4-6. Magnetic field lines in front of the gap in a conventional recording head. lg is the gap length, and the numbers 1, .9, .8 etc., indicate the relative field strength.

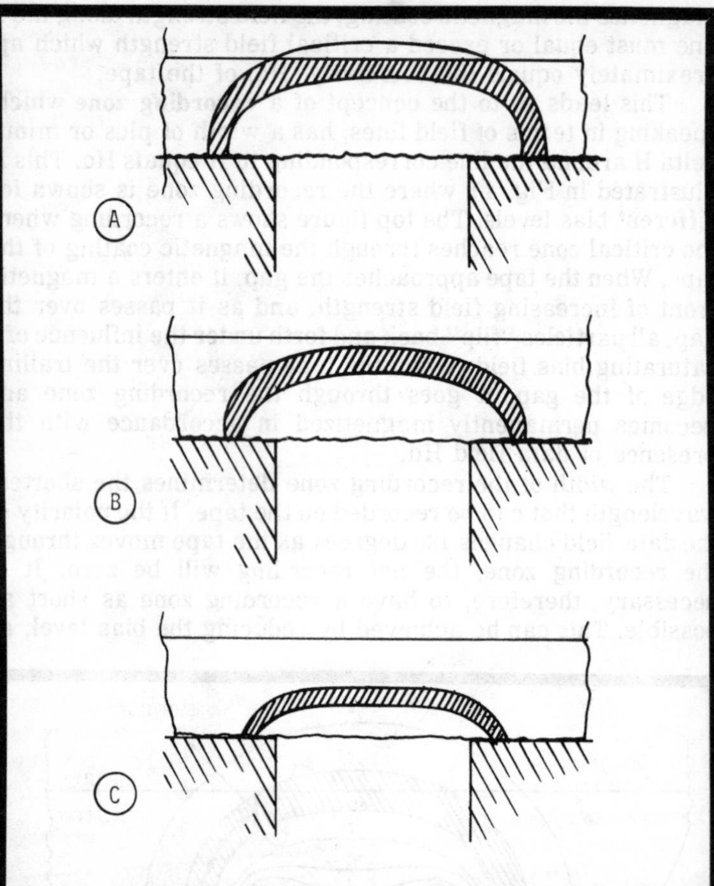

Fig. 4-7. These drawings indicate the recording zones present in front of a record head: high bias level (A), low level (C), and a normal level (B).

shown successively through Fig. 4-7A, B, C. But as we reduce the bias level, the tape is no longer recorded in its full thickness and the level at long wavelengths is reduced and the signal highly distorted.

During the past few years considerable effort has been devoted to the design of record heads with a narrow recording zone[2]. Other constructions use eddy current phenomena for field-shaping[3,4]. The conventional ring core head has an optimum length of .25 mils (6 mu) and a recording zone with a width of approximately .030 mils (.8 mu). This limits the shortest wavelength to about .060 mils. The newer heads have recording zones that are substantially narrower, and recordings of wavelengths down to .010 mils may become feasible (100,000 cycles/inch). But the overall recorder/reproducer response will still be limited by other losses, such as the coating thickness loss, the spacing loss, and the reproduce head gap loss, which are discussed later in this chapter.

REMANENT MAGNETIZATION ON THE TAPE

The external flux lines from the recorded tape determine the voltage that will be induced in the reproduce head. Fig. 4-8 shows the internal magnetization of the individual particles after leaving the recording zone. (Non-oriented small particles are assumed in Fig. 4-8.) The influence of particle orientation is considered later in this chapter. At long wavelengths, all the magnetized particles contribute to the external flux which, therefore, is proportional to the coating thickness and remanence Br of the tape.

As the wavelength decreases (1/lambda increases), fewer and fewer of the magnetized particles contribute to the external flux which, therefore, decreases. This loss has often been called **self-demagnetization**, which is a somewhat incorrect name. Each particle is a minute permanent magnet with either plus Br or minus Br remanence. The internal flux lines, as shown in the right-hand portion of Fig. 4-8, serve as keepers for neighboring particles. Only the particles in the tape surface contribute to the external flux at short wavelengths and the thinking in terms of demagnetization, like in bar magnets, is questionable.

This presentation also gives a natural explanation of the 6 db/octave loss in external flux when the wavelength becomes shorter than 2 pi c, where c is the coating thickness. The recording in depth was originally presented as a "bubble" theory [5], although it was a weak model by virtue of the fact that it considered only the longitudinal field component of the recording field. The theory is at its best when only tapes with acircular (elongated) longitudinal-oriented particles are considered at long wavelengths. In a small-particle, non-oriented tape, each particle becomes magnetized when it passes through the recording zone and is oriented in accordance with the recording field lines. And since this direction is predominantly controlled by the AC bias field (which is 8 to 10 times greater than the data field) it becomes obvious that the remanence at very short wavelengths is predominantly perpendicular. This phase shift has been observed in instrumentation recorders, as is described in Chapter 5.

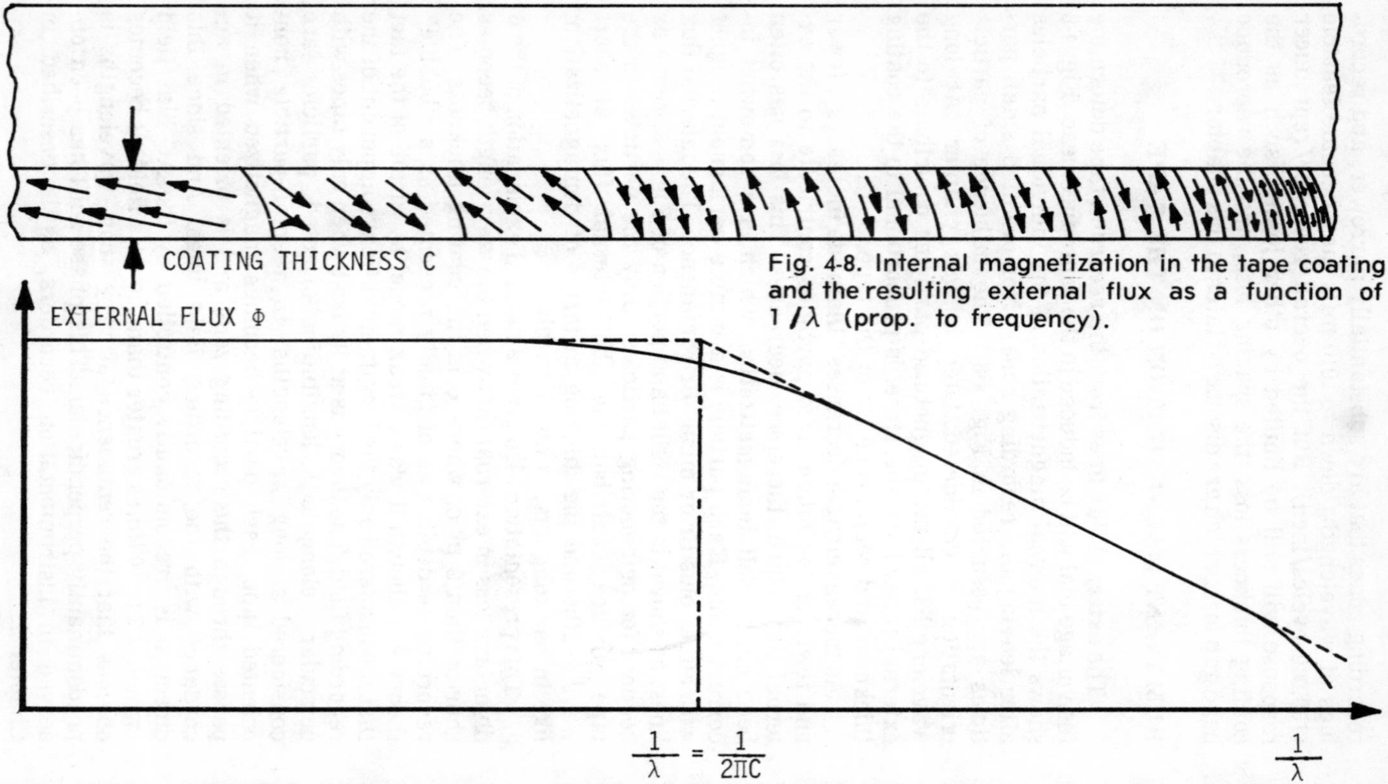

Fig. 4-8. Internal magnetization in the tape coating and the resulting external flux as a function of $1/\lambda$ (prop. to frequency).

The loss in external flux can, therefore, most correctly be called a thickness loss, expressed as:

$$\text{Thickness Loss} = 20 \log \frac{2\pi/\lambda}{1 - e^{-j}\frac{2\pi c}{\lambda}}$$

where minus j is the phase shift in remanence.
The combined loss due to a reduction in external flux and the final width of the recording zone is shown in Fig. 4-9. Curve A represents tape flux ϕt with bias adjusted for optimum level at long wavelengths. As the bias level is decreased, the long-wavelength flux also decreases but the short-wavelength flux increases due to the narrower recording zone (as shown in Fig. 4-7).

PLAYBACK

When the recorded tape passes over the last head (the reproduce head in Fig. 4-1), the external flux threads through the core and induces voltage Vplay in the winding. If the core has n turns in the winding and the flux is:

$$\phi = \phi c \cdot \cos\omega t$$

where: $\omega = 2\pi f$

then the induced voltage is:

$$Vplay = - n \frac{d\phi}{dt}$$

$$= n.\omega.\phi c.\sin\omega t$$

This equation expresses that the output voltage is proportional to the number of turns in the winding, frequency f and flux ϕc through the core.

The induced voltage is shown in Fig. 4-10A as a differentiation of curve A from Fig. 4-9. When measured, the curve is found to be somewhat lower in level. This is due to losses in the reproduce process:

Core loss
Gap length loss
Spacing loss
Alignment loss

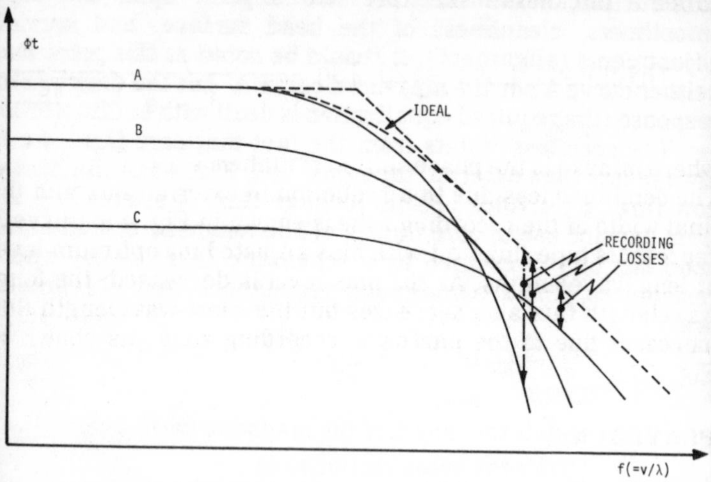

Fig. 4-9. The net resulting external tape flux Φt as a function of frequency (or 1). Curves A through C represent high through low bias levels.

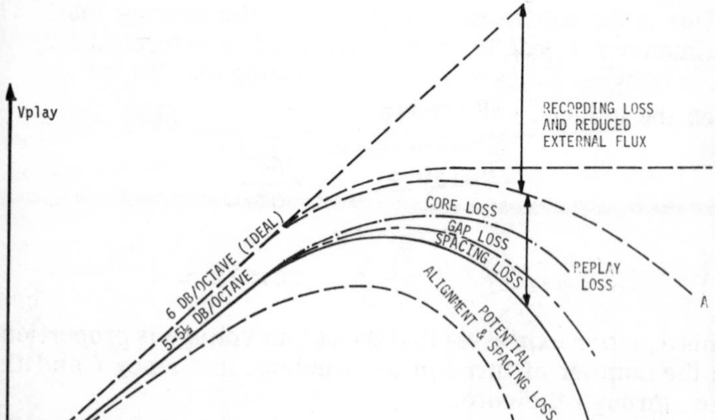

Fig. 4-10. Replay voltage vs frequency. Curve A is the induced voltage, as calculated, while A′ is the measured voltage.

The first two depend upon the reproduce head design and construction, while the last two depend upon the tape smoothness, cleanliness of the head surface, and service adjustments (alignment). It should be noted at this point that neither curve A nor the measured curve A' has the desired flat response (the required equalization is dealt with in Chapter 5).

The core loss results from the fact that core flux ϕc is always less than external tape flux ϕt. Fig. 4-11A illustrates how flux ϕt divides up into leakage flux ϕs, crossing the front gap and useful core flux ϕc. The magnetic circuit of the head can be represented by a schematic (Fig. 4-11B) where resistances Rm are magnetic resistances that universally are:

$$Rm = \frac{L}{\mu o \mu r A}$$

where: L equals the length of the magnetic path

A equals the cross section area

μr equals the relative permeability

μo equals $4 \pi \times 10^{-7}$ Henry/meter.

Rmfg represents the magnetic resistance of the front gap, Rmbg the back gap and Rmc the resistance of one core half (Rms is the equivalent resistance for the spacing loss). The efficiency (η) of the reproduce head is determined by the ratio between Rmfg and 2Rmc plus Rmbg plus Rmfg:

$$\eta = \frac{Rmfg}{2Rmc + Rmbg + Rmfg} \cdot 100\ \%$$

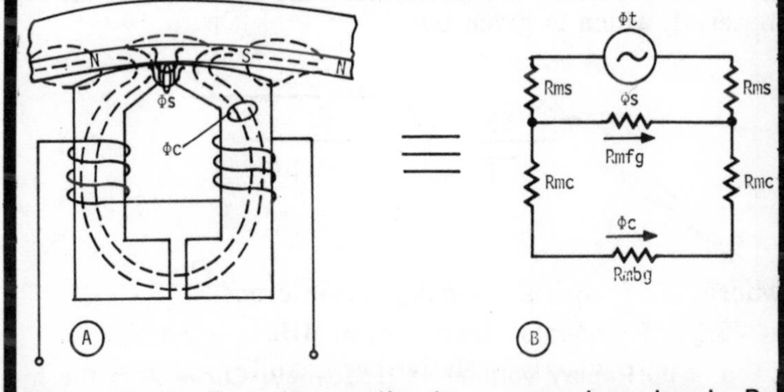

Fig. 4-11. A. Flux line paths in a reproduce head. B. Equivalent schematic of a magnetic head.

The value of η is typically around 90 percent for audio heads but can drop to 50 percent for high resolution heads, which have a very narrow gap (small Rmfg). At very high frequencies, it can drop 15 to 20 percent due to eddy current losses which increase Rmc. We can now substitute ϕt times η for ϕc:

$$Vplay = n.\omega.\phi t.\eta.\sin\omega t$$

The reduced voltage due to the efficiency factor is indicated in Fig. 4-10.

The upward slope of the output voltage should ideally increase at a rate of 6 db/octave, but it is not unusual to measure a slope of 5 to 5.5 db/octave. This is largely due to a decrease in efficiency because Rmc increases with frequency due to eddy current losses. These losses are in practice reduced by fabricating the head cores from thin laminated mu-metal or ferrite. It is appropriate to mention here that core losses quite often limit how high a bias frequency can be used in the record electronics. Too high a bias frequency may require excessive bias current to overcome the eddy current losses, and the hysteresis losses may cause the record head to heat. (The hysteresis losses are proportional to the product of the frequency and the area of the hysteresis loop.)

In practice the eddy current losses can best be visualized as a reduction in the cross section area A. The eddy currents limit the penetration of flux lines into the core center, and they are confined to the core surface. As the frequency increases, this surface sheath is no thicker than the **skin depth** of the core material, which is given by:

$$\delta = \frac{66}{\sqrt{f}} \sqrt{\frac{\zeta}{\zeta c.\mu r}}$$

where: δ equals skin depth in microns
f equals frequency in MHz
μr equals relative permeability
ζ equals resistivity of core material
ζc equals resistivity of copper.

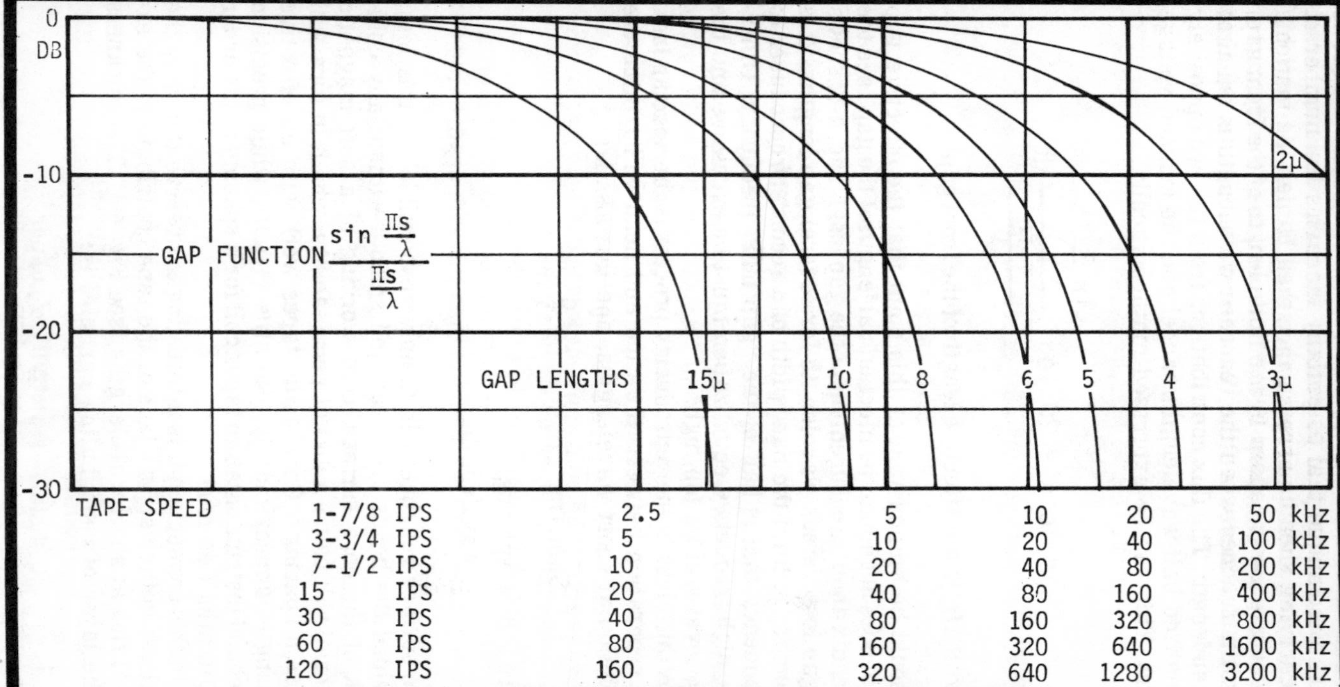

Fig. 4-12. Gap loss as function of frequency for various gap lengths.

A laminated structure, therefore, is used in most magnetic heads, a technique that essentially increases the number of flux-carrying sheaths (two surface sheaths per lamination).

The gap loss is caused by the finite length of the front gap. A limit is reached when the recorded wavelength is equal to the gap length. The flux contributions from the two oppositely magnetized half-wavelengths cancel and the induced voltage is zero. This can be expressed mathematically:

$$\text{Gap loss} = 20 \log \frac{\sin \frac{\pi l_{fg}'}{\lambda}}{\frac{\pi l_{fg}'}{\lambda}} \text{ db}$$

where: l_{fg}' is the effective length of the front gap

It should be emphasized at this point that the effective gap length does not equal the mechanical length of the gap (see the Appendix about measurement of the gap loss). Fig. 4-12 shows the gap loss versus wavelength for several gap lengths. It is customary to limit the bandwidth of a recorder/reproducer to a frequency that is half of the "gap null" frequency. Otherwise, high-frequency equalization must be excessive and the noise level will be too high.

A third loss in the reproducing process is the **spacing loss.** Any separation between the tape and head will reduce the resolution at short wavelengths, and this loss is:

$$\text{Spacing Loss} \simeq 55 \cdot \frac{d}{\lambda} \text{ db}$$

where: d = spacing
λ = wavelength.

Tape surface roughness is mainly responsible for this loss, provided the heads are clean and smooth without any cold-work of the core surface (cold working of a soft magnetic material will cause a loss of permeability, which in turn will act as a spacing). Good audio tapes work with an effective spacing of approximately .040 mils (1 mu), while precision wideband instrumentation tapes call for a spacing in the order of .010 mils (.25 mu).

In this connection it is interesting to observe that only the particles in the very surface of the tape contribute to the external flux at short wavelengths (see Fig. 4-8). The thickness of the layer of contributing particles is:

$$y_o = .22 \cdot \lambda$$

At a frequency of 15 kHz at 1⅞ IPS (1 MHz at 120 IPS) this corresponds to .22 x .120 or 0.26 mils, which is comparable to the particle size.[7]

Further, if the reproduce head gap is tilted in relation to the recorded pattern, then an additional loss occurs. This is the alignment loss, and is:

$$\text{Alignment Loss} = 20 \log \frac{\sin \frac{\pi s'}{\lambda}}{\frac{\pi s'}{\lambda}} \text{ db}$$

where: $s' = w.\text{tg}\,\alpha$

w = track width

α = misalignment angle

Most recorders have their reproduce heads mounted on a plate that allows for correct adjustment of the alignment angle. It should be noted that the alignment is less critical when the track width is narrow.

PULSE RECORDING WITHOUT AC BIAS

The pulse recording technique is used almost exclusively in computer installations where the main performance criteria is the presence or absence of a signal, resulting in a series of pulses. This recording technique is simple and is used with the binary number system, where the presence or absence of a signal may represent a 1 or a 0. The recording is made by simply switching the record current (DC) on or off, or by reversing its polarity. The first method is known as RZ (**Return to Zero**), the latter NRZ (**No Return to Zero**). There are other schemes for pulse recording, but these are generally made with an AC bias, and the record-playback functions and quality are similar to what was described earlier. (Some of these techniques are mentioned in Chapter 9.)

Although pulse recording with AC bias appears rather simple, its performance is somewhat limited by the so-called peak-shift during the playback process. For example, when the pulse current through the recording head is turned on or off instantaneously, it effectively snaps the external flux field of the recording head onto the tape.[8] That is, the sudden current change leaves in the tape a "snapshot" of the instantaneous flux distribution. This is illustrated in Fig. 4-13, where a few RZ pulses are recorded (on-off-on-off). Shown to the left is the record-head field and the magnetization patterns in the tape are shown to the right.

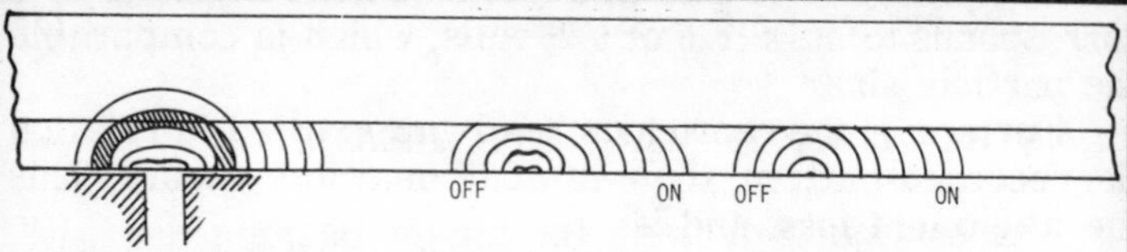

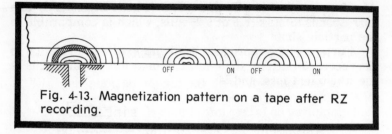

Fig. 4-13. Magnetization pattern on a tape after RZ recording.

When a digital tape is played back, the flux reversals induce voltage spikes in the reproduce head winding (a voltage is induced only during flux changes $d\phi/dt$) and two observations can be made from Fig. 4-13. First, it is noted that the pulse fronts are not identical and when the pulses approach the reproduce head and contact the gap, it is obvious that the induced voltages are different for the two "magnetization" fronts. Second, although the current was turned on at regular time intervals, the fronts are not spaced equally. Both factors limit how close pulses can be recorded (packing density). It has been determined that the ability of the reproduce head to define the position of the pulse on the tape is only one third to one sixth times as effective as that of the record head in placing the signal pulse there.

Other factors affect the performance of pulse recordings, in particular the thickness of the coating and the spacing of the head from the tape surface. (A more proper word here would be media surface, since non-contact recording and playback are applied in computer discs. See Chapter 9.) The pulse amplitude is essentially proportional to the coating thickness, while the packing density is inversely proportional to the one-half power of the thickness. Other factors also play a role in digital recording performance, such as the coercivity and remanence of the coating. (The correlation of these properties and pulse recording is still under heavy investigation by several researchers.)

Spacing between the pulse record/reproduce head and the coating causes both a reduction in output amplitude and packing density. This type of spacing loss can cause error in a digital recorder as a result of "**dropouts**" in the tape (or media) surface. Dropouts are momentary losses of signal and are caused by either holes in the tape surface or foreign particles which lift the tape away from the head.

It should be noted, finally, that pulse recordings in the NRZ mode often are done on a media that has a previous recording on it. The recording field is strong enough to saturate the coating, but any previous recording will act as a

DC bias, influencing the position of the magnetization front when the record field is switched.

CONSTRUCTION OF MAGNETIC HEADS

The continued demand for higher packing densities on magnetic tapes helps keep the pace of head development and manufacturing up to the state of the art. The ring core size has decreased from about one-inch diameter to the size of the end of a matchstick. This is necessary to reduce eddy current losses in high-frequency recording and playback. At the same time, the track width has been reduced from 6.25 millimeters (250 mils–full-track recording on ¼" magnetic tape) to .5 millimeter (20 to 25 mils–eight-track recording on ¼" tape). The limitations for further reductions in track width are in the areas of cross-talk and tape guidance.

Some typical magnetic heads are shown in Fig. 4-14A. Although they appear quite similar, except for size and track arrangement, they are designed to meet different criteria depending on their use as erase, record or reproduce heads. The differences are indicated in Fig 4-1 where the **erase head** has a fairly large core, a wide backgap, and a spacer in the front gap. The spacer in the front gap acts as an eddy current shield which tends to force the flux from the core out into the tape coating. The large backgap is essential to keep the erase head free of any permanent DC magnetization which can occur when the erase current is switched off. The field from the erase head must be capable of saturating the magnetic tape; therefore, a rather large erase current, is required. This results in heat generation due to hysteresis losses. Ferrites are often used in erase heads, for the ferrites have a high electrical resistance and, consequently, small hysteresis losses. However, the ferrites are difficult to use in magnetic heads because of their mechanical weakness.

Erase heads are used almost exclusively in home recorders (and some professional recorders). It is common practice to use larger bulk degaussers in instrumentation and computer systems whereby an entire reel of tape can be erased. Various bulk degaussers or demagnetizers, as they are also called, are covered in Chapter 6.

Referring again to Fig. 4-1, the **record** and **reproduce heads** are similar, but differ in core size. The reproduce core is normally made as small as practicable in order to reduce eddy current losses. This is of less importance in the record head, since additional current to overcome losses is provided by the design of the record amplifier.

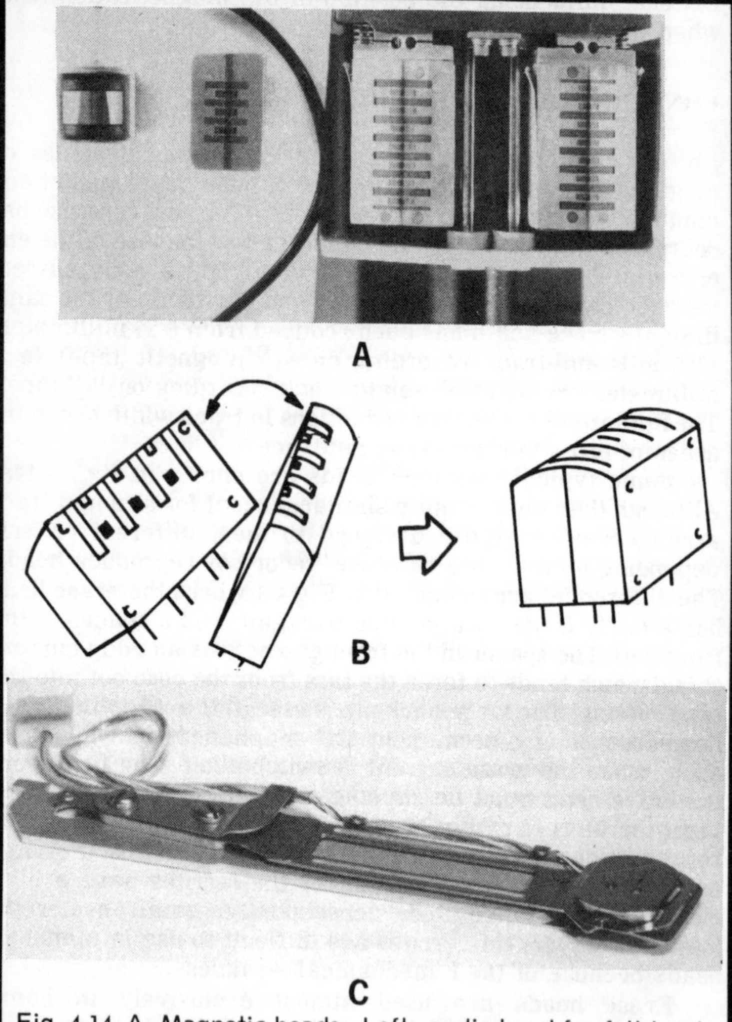

Fig. 4-14. A. Magnetic heads: Left: audio head for full-track recording of ¼″ tape; middle: three-channel head for ½″ tape; right: two interleaved seven-channel instrumentation heads for recording fourteen tracks on a 1″ tape. B. Assembly of a magnetic head. Flying head assembly (C) for disc memories. The pad to the left contains the write-read core. When the head is positioned over a rotating disc it will actually fly 100-200 micro-inches over the disc due to the air stream generated by the rotation and the aerodynamic design of the head pack. (Courtesy Applied Magnetics Corp., Goleta, Calif.)

Fig. 4-14B illustrates how the typical magnetic head is fabricated.The core itself, which most commonly has the form of a C, is either stamped or etched from thin sheets of a material that has a high degree of permeability. It is then carefully annealed and the laminations are stacked, with insulators between, to the desired track width. After winding the coils on the core halves, they are inserted in a so-called half-shell and kept in place with an epoxy cement.

Now comes the most difficult process in head manufacturing; namely, the lapping of the two head halves to provide for perfect plane mating. The lapping process and the compound used must be carefully selected, since it is essential that any high points in the core should be cut off rather than compressed to the point of flatness. If the ends of the core at the gaps become cold-worked during the lapping process, the permeability becomes lower and this will in effect act as a longer gap than the one resulting from a mechanical or optical measurement of the finished gap lengths. The final steps in the lapping process require a very high degree of skill, and often the attainment of the desired plane surfaces depends more upon craftsmanship than the carefully prescribed process control.

The lapped head halves are now placed against each other and the gap length is controlled either by the insertion of non-magnetic materials such as a metallic foil or by vacuum deposition of a spacer; for example, silicon-monoxide. The last process is probably the most commonly used, in particular for microgap heads where the gap lengths are in the order of 30-40 microinches. Such tiny gap lengths would be impossible to provide with a metallic spacer. Modern vacuum techniques also allow for deposition of metal films for spacers, which alters the head's electrical performance, particularly at high frequencies, by increasing its efficiency.

The two head halves are held together either by screws, pins or by the use of an epoxy cement and the assembled head block is now ready for its final processing. The front of the head is ground down to the desired curvature, then it undergoes a careful lapping to provide for a mirror-smooth surface finish. This again is a critical process that requires great care and skill.

The core material is generally a highly permeable material such as High Mu 80 or Permalloy. These have the disadvantage that while they are soft magnetically with a high permeability, they are also soft mechanically and subject to fast wear. In addition, there is the potential danger of gap smearing whereby the material cold-flows across the gap and

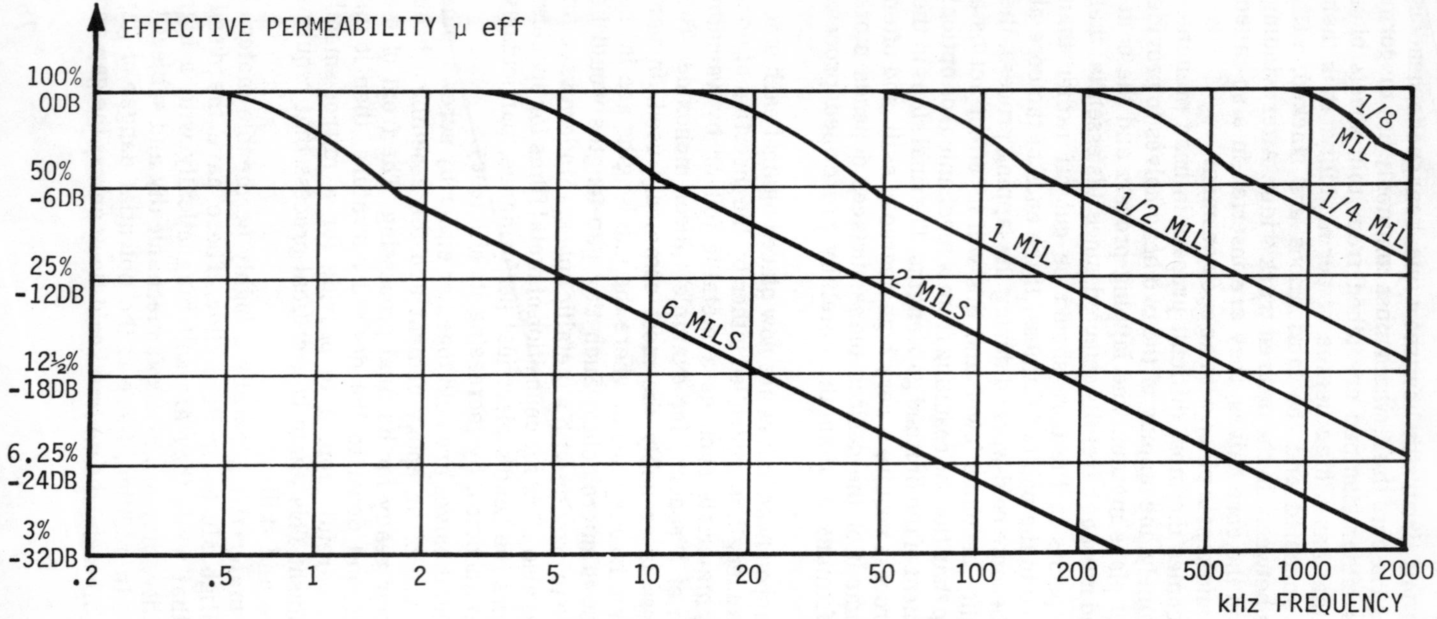

Fig. 4-15. Reduction of permeability (μ) with increased frequency and at various lamination thicknesses. The lamination thickness affects the efficiency of the magnetic heads at high frequencies.

in essence short-circuits the magnetic flux from the tape as it enters the core. New materials with a somewhat lower permeability, but with ten times the hardness, are now available in an alloy of iron and aluminum, generally referred to as Alfenol but having different trade names (Alfesil, Vacudor). It is common that all such cores be laminated to avoid excessive high-frequency losses. Fig. 4-15 illustrates how the effective permeability is reduced by a frequency increase for various lamination thicknesses. For any wideband use, such as recorders extending to one or two MHz, it is necessary, therefore, to use laminations in the thickness range of 1 to 2 mils. Since such thin laminations are very difficult to handle, other ways are often sought to reduce the high-frequency losses. Most commonly used are ferrites, which are provided with pole shoes of say, Alfenol, to obtain clean and straight gap edges (Fig. 4-16). In mechanical structure ferrites are a ceramic and it is exceedingly difficult to lap the gap edges to provide a straight gap, which is essential for high-quality recording and playback.

The core geometry and the selection of material are determined by the desired operating range of the particular head, whether it is for recording or playback. The geometrical shape of the **front gap** is an important design factor[9]. Fig. 4-17 shows three different front-gap shapes. Fig. 4-17A shows generally pointed pole tips, which provide for high sensitivity since the magnetic resistance of the front gap is high. However, it is extremely wear-sensitive and is, therefore,

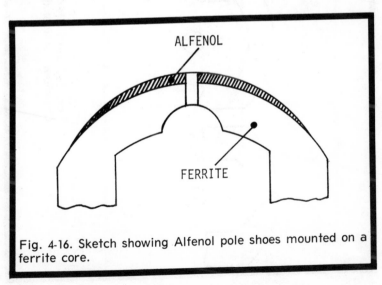

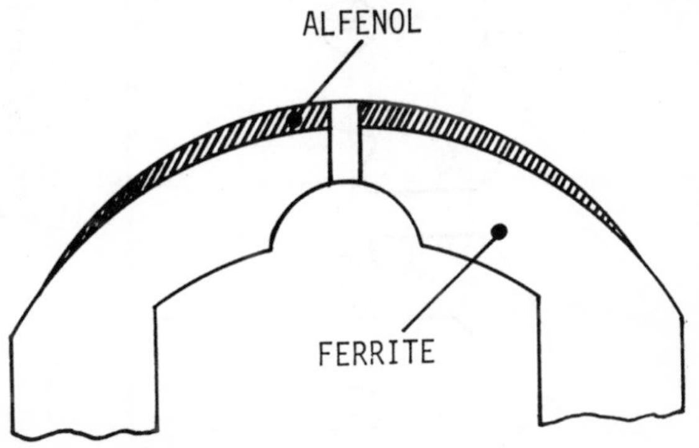

Fig. 4-16. Sketch showing Alfenol pole shoes mounted on a ferrite core.

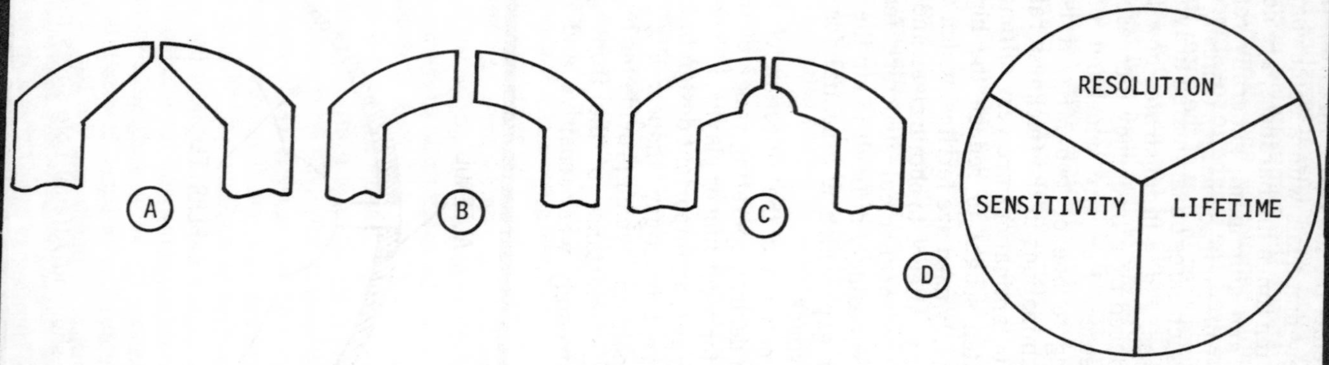

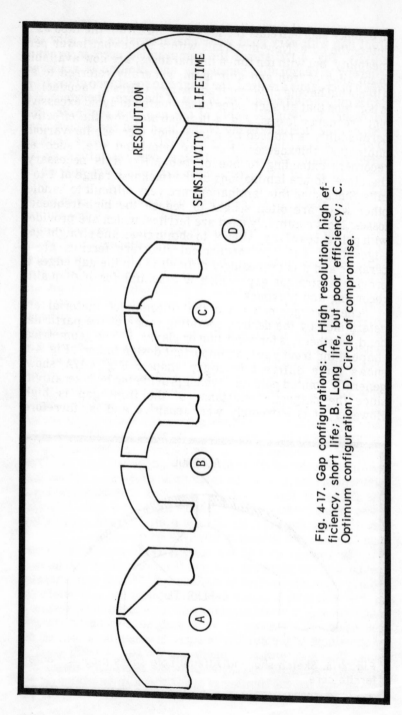

Fig. 4-17. Gap configurations: A. High resolution, high efficiency, short life; B. Long life, but poor efficiency; C. Optimum configuration; D. Circle of compromise.

never used in the practical application. Fig. 4-17B shows a front gap with very good wear characteristics but poor sensitivity. Fig. 4-17C shows the compromise used in practice whereby a reasonable sensitivity and lifetime is achieved. When microgaps are used, the magnetic resistance of the front gap is quite small compared to that of the core; therefore, it is necessary to limit the depth of the gap to a few mils. This reduces the useful lifetime of the head and the design compromise is best illustrated (Fig. 4-17D) by a circle showing three sections: sensitivity, lifetime, and resolution. An increase in any of the three sectors can take place only at the cost of one or both of the others.

Multi-track heads are in wide use today, such as in stereo recording and playback, computers, and instrumentation. Special care must be exercised in the design and construction of these heads to avoid crosstalk. Two adjacent cores in the same head can interact with each other at low frequencies, as in a transformer. At high frequencies the capacitance between windings may cause transfer of signal information from one core to the other. The prevention of crosstalk thus requires careful shielding between cores and it has become common practice to use interlaced heads, where two seven-channel heads are used for recording or playing back 14 tracks on a 1" wide instrumentation tape. The tracks in this case are each 50 mils wide and the head shields are also designed to be 50 mils wide, so, in addition to shielding, they serve as a shunt for the flux in the tape. The transformer-action flux and the fringing flux from adjacent tracks are essentially the limiting factors for crosstalk at low frequencies. This is very pronounced in the so-called **flux-sensitive heads**, where a semi-conductor element is inserted either in the front gap or the back gap. This element is sensitive to the magnitude of the flux rather than the change of flux ($d\phi/dt$). Although flux-sensitive heads (also called "Hall element" heads) are useful down to a very few Hz, they are severely limited in multi-track applications due to excessive crosstalk at these frequencies.

The **impedance** and the **current requirements** for record heads, and the **playback voltages** from reproduce heads, vary widely from application to application. This is demonstrated in Fig. 4-18 which shows typical impedances and levels for audio recorders at one extreme and for wideband recorders at the other. The sensitivity of a magnetic head is directly proportional to the number of turns that can be wound on the core. The distributed capacitance in the winding—and in particular the input capacitance to the reproduce amplifier—sets a limit for the number of turns that can be wound on the

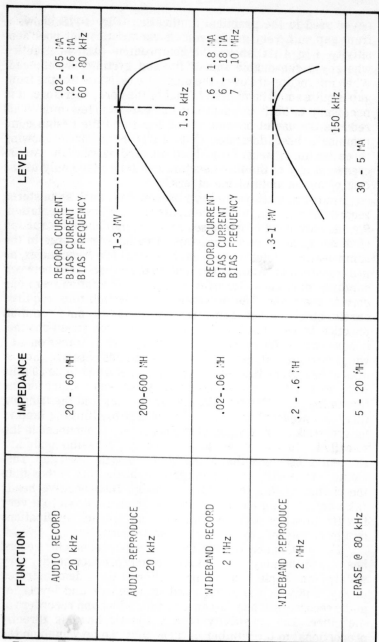

FUNCTION	IMPEDANCE	LEVEL
AUDIO RECORD 20 kHz	20 - 60 MH	RECORD CURRENT .02-.05 MA BIAS CURRENT .2 - .6 MA BIAS FREQUENCY 60 - 80 kHz
AUDIO REPRODUCE 20 kHz	200-600 MH	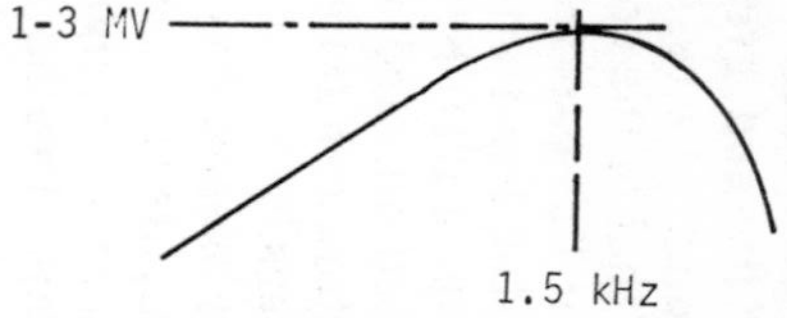
WIDEBAND RECORD 2 MHz	.02-.06 MH	RECORD CURRENT .6 - 1.8 MA BIAS CURRENT 6 - 18 MA BIAS FREQUENCY 7 - 10 MHz
WIDEBAND REPRODUCE 2 MHz	.2 - .6 MH	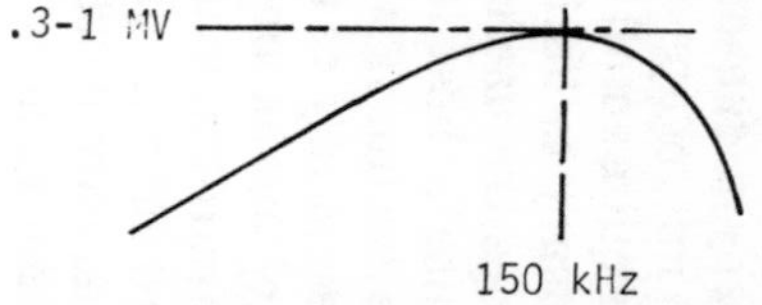
ERASE @ 80 kHz	5 - 20 MH	30 - 5 MA

Fig. 4-18. Table of magnetic head nominal electrical impedances and input and output levels for recording and reproducing a 1.25mm (50 mils) wide-track on standard magnetic tape.

core. The resonant frequency for a record head must be higher than the bias frequency used and the resonant frequency for the reproduce head must be higher than the highest frequency to be reproduced. Examples of this are given in Chapter 5.

MANUFACTURE OF MAGNETIC TAPE

Some twenty firms in the U.S.A. and about a dozen in Europe are presently engaged in manufacturing all types of magnetic tape. Although there may be subtle differences between the different brands and different types of tapes, they are basically of similar construction. The detailed chemical formulations and processes are trade secrets, but a general outline of magnetic tape materials and manufacturing is given below.

A modern general purpose magnetic tape consists of a one-mil thick Mylar film with a coating consisting of magnetic particles in a binder, with a thickness of from .2 to .4 mils. Fig. 4-19 shows photographs of magnetic particles and a cross-section of a magnetic tape as viewed through an electron microscope. The coating is seen to consist of a binder material with a large number of magnetic particles. These particles are

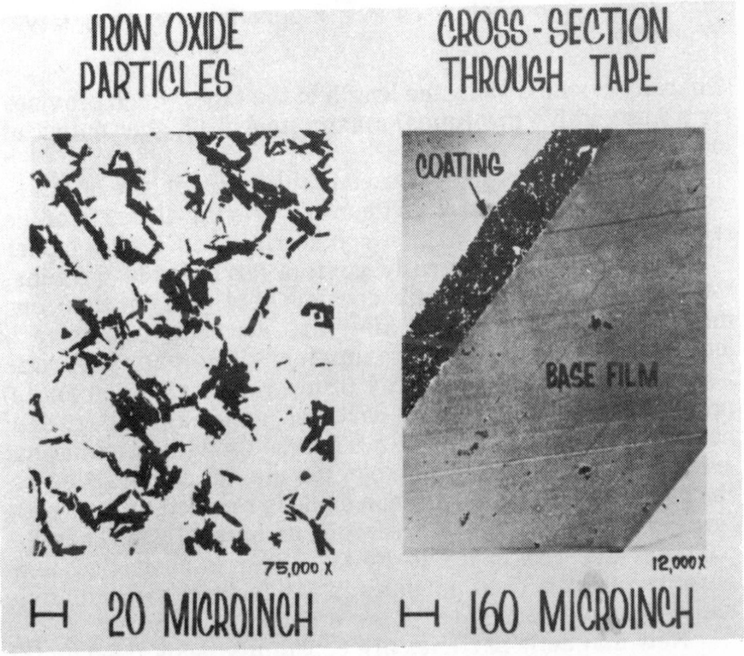

Fig. 4-19. Photo of iron oxide particles and a cross-section of a magnetic tape. (Courtesy Memorex Corp.)

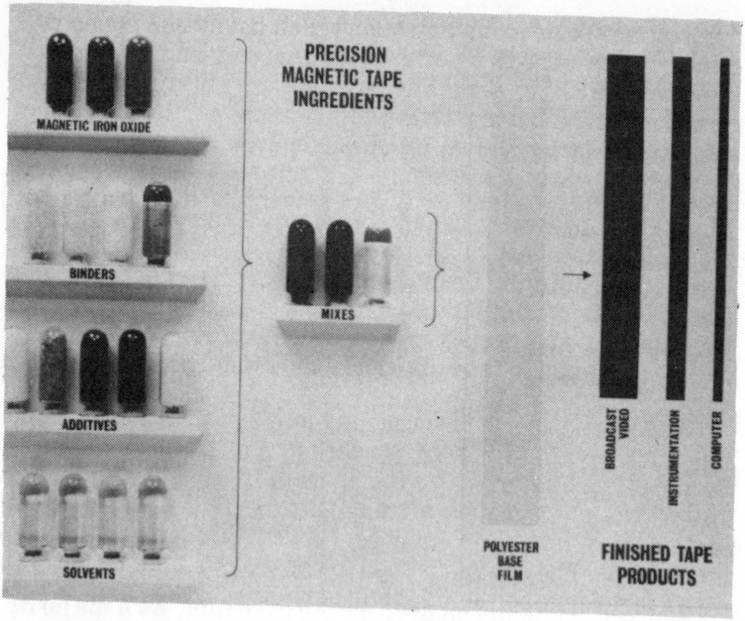

Fig. 4-20. Components used in a magnetic tape. (Courtesy Memorex Corp.)

generally oriented along the length of the tape, which provides for a 3 db (with a theoretical maximum 6 db) higher output at long wavelengths.

Orientation is a process used in the manufacture of tapes, causing the particles to align themselves in the direction of the recording field. The **magnetic material** used in the binder system is almost universally gamma-ferric oxide (gamma-Fe_2O_3). These particles are considerably shorter than one micron (40 microinches); they are acircular and have a length-to-width ratio of approximately six to one. The oxide particles are manufactured by first dissolving iron in an appropriate acid. The resulting particles, called alpha particles, have the elongated shape of the finished magnetic oxide but are non-magnetic. To continue in the manufacturing process, the particles are heated and the oxide is reduced to the Fe_3O_4 form in an atmosphere of hydrogen or natural gas. Then the particles are reoxidized to Fe_2O_3 under a controlled temperature and the final particles in the form of gamma-ferric oxide are obtained.

New magnetic particles are constantly being tried by the various tape manufactures in order to improve the overall performance of magnetic tapes. Among these new materials are gamma-ferric oxide particles which have been doped with cobalt; these particles suffer from thermal and mechanical instability with a resultant erasure of short wavelength recordings. The basic gamma-ferric oxide particles are now available in large and small particle sizes. An advantage of smaller particles is a lower noise tape, essentially because there are more particles per wavelength than with conventional oxides. Another promising material is chromium dioxide which is an elongated particle like the iron oxide. It requires extremely accurate control in manufacturing and distinguishes itself from the iron oxides by having a higher coercivity, a higher squareness ratio, and a lower Curie temperature. The higher coercivity force may present a problem in present-day equipment since it requires a higher bias current, which may lead to poletip saturation, but this present disadvantage may be overcome in the future by the use of newer record-head construction[4]. The practical aspect of a lower Curie temperature has not yet been fully evaluated.

The ingredients for a magnetic tape are shown in Fig. 4-20. The **chemical formulation** of the binder material depends largely upon the material used for the base film. As a majority of tapes today have plastic bases, we find that the binder dispersions usually contain the following components:

Binder material: In essence the cement which holds the magnetic particles together when the dispersion has been applied to the base and the solvents have dried. The generally used binder today is one of the bakelite compounds; that is, vinylite.

Plasticizer: This provides flexibility to the binder material.

Wetting agent: Since the magnetic particles used in the coating are magnetic, they agglomerate in lumps, and it is the purpose of the wetting agent to break up such agglomerates and provide for the best possible dispersion of the particles. Examples of agglomeration are shown in Fig. 4-21. Such agglomerates deteriorate the signal-to-noise ratio of a recording.

Resin: The function of the resin is to toughen the dried coating; too much will cause the coating to be sticky, while too small an amount will cause the coating to be brittle and tend to flake.

Solvents: Two or more solvents are generally included in dispersions; first, to provide a thinner for dispersion and

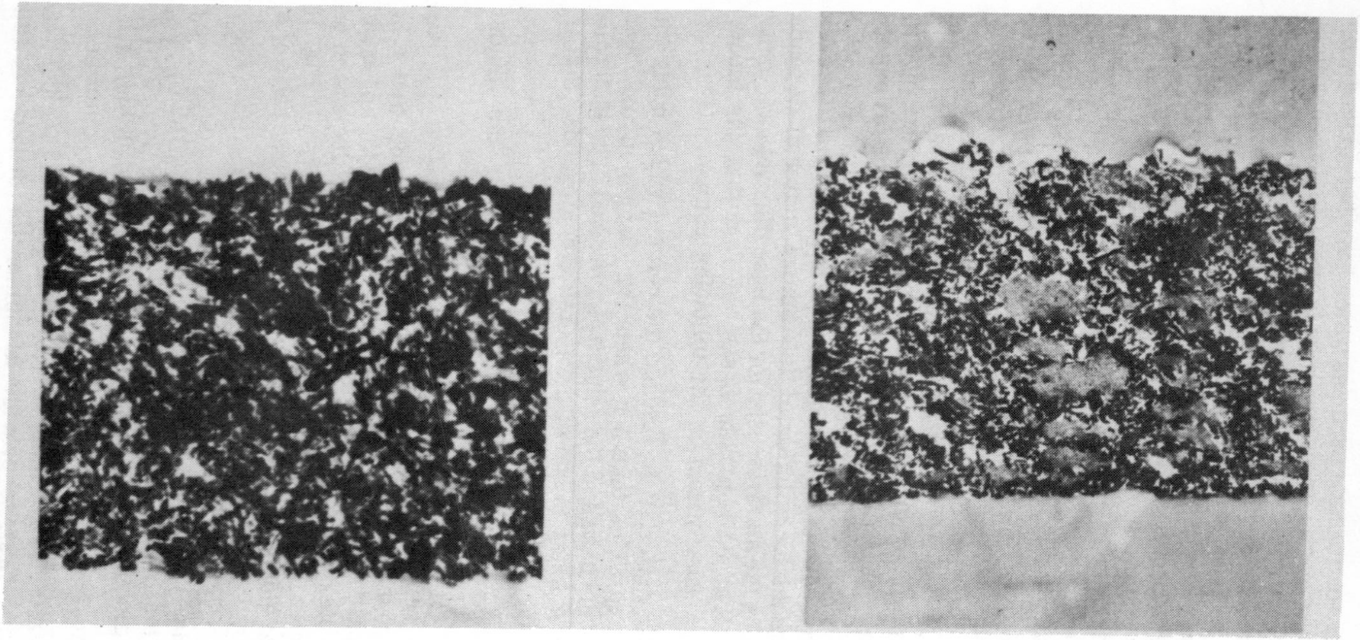

A

B

Fig. 4-21. Electron microscope photos of a well dispersed coating (A) and a poorly dispersed coating (B), showing agglomerations and voids. (Courtesy Memorex Corp.)

second, to provide a good bond between the binder and coating.

Lubricant: To overcome stickiness and scrape-flutter problems, a lubricant is generally added to the binder material. Great care should be exerted in the type selection and the amount applied, since it may transfer to the backside of the base material, which would result in poor friction characteristics between the capstan and the rubber puck.

Other ingredients may be added to the dispersion, depending upon the particular use. As an example, anti-fungus agents are added to dispersions for tapes for military use.

Three types of material are used for the **magnetic tape base**: celulose acetate film, polyvinyl chloride plastic, and polyester. These film materials have greatly varying physical properties, and since these are associated with the practical operation of a recorder they are discussed in Chapter 6 which is devoted to the selection of tapes and accessories.

The process of **manufacturing** a magnetic tape starts, initially, with the milling of the magnetic particles into the binder, also called the dispersion process [10]. The function of this process is to distribute the oxide particles uniformly throughout the binder. The oldest and most common dispersion method is a ball mill, which is simply a jar or a bell which is partially filled with a dispersing media, either metallic or ceramic in the shape of balls, cylinders, rods, or of a random shape such as in sand. The time of the milling process may last from a few hours to several days, and it is dependent upon the chemical composition of the binder and the particles used.

After the milling process, the material is fed to a coater which applies the coating onto the film. Several coating processes are used, such as a knife coater, the reverse roll coater, and the gravure coater. See Fig. 4-22A, B, and C. Immediately after coating, while the binder is still wet, the coated web undergoes orientation, whereby the magnetic particles are aligned. This provides, as mentioned earlier, for increased output at long wavelengths. The effectiveness of orientation is generally referred to as **squareness ratio**, which is the ratio between the remanent saturation induction and the saturation induction. For randomly-oriented particles, the squareness ratio is 0.5, and for ideally and perfectly oriented particles the ratio is equal to 1.0, providing for a 6 db increase in long wavelength output. Practical values for the squareness ratio normally falls around .7, which gives an increase of 3 db.

After coating and orienting, the tape enters a drying oven,

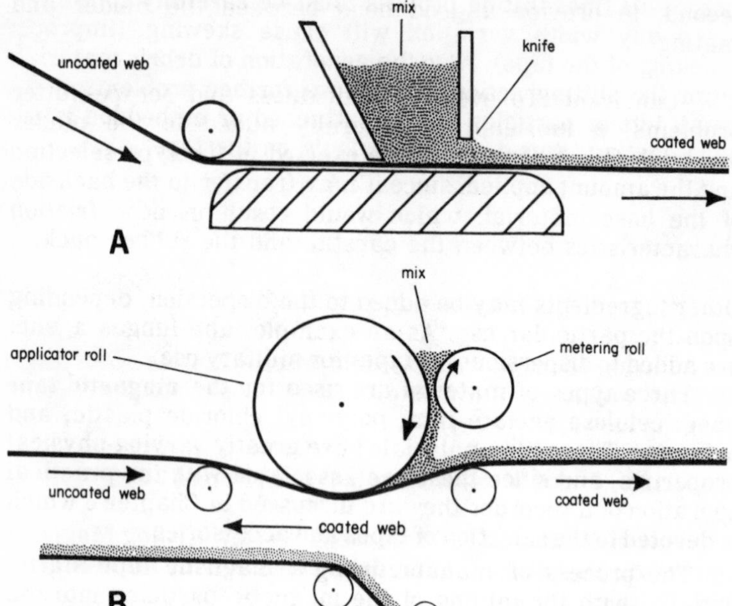

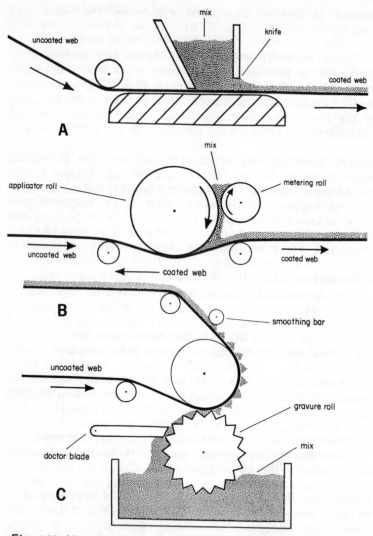

Fig. 4-22. Magnetic tape coating methods: A. knife coater; B. reverse roll coater; C. gravure coater. (Courtesy Memorex Corp.)

which normally is several hundred feet long, where all the solvents are evaporated, leaving a coated web of tape. The tape is wound onto a large roll which varies in width from 6 to 24 inches. The finished web now undergoes a surface treatment which depends on the intended use of the tape. After this treatment, the web is slit into the finished widths, which may be from one-fourth inch for audio uses up to 2 inches for video uses. This final slitting process must be carefully controlled, since any width variation will cause skewing (improper tracking of the tape). Also the generation of debris materials from the slitting action may cause further problems if the small bits or particles become attached or embedded in the tape coating, thus causing dropouts on the tape.

References:

1. Neel, L., Cahiers de Physic, 1943, pp. 17-47.

2. Camras, Marvin: "An X-Field Micro-Gap Head for High Density Magnetic Recording," IEEE Conv. Record, Pt. 6, 1964, pp. 359-371.

3. Went, J. J. and W. K. Westmijze: U. S. Patent 2,854,524.

4. Johnson, W. R. and F. Jorgensen: "A New Analog Magnetic Recording Technique," Proc. Intl. Telemetering Conferences, Vol. II, 1966, pp. 414-430.

5. Mee, C. D.: "The Physics of Magnetic Recording," North-Holland Publishing Company, Amsterdam, 1964.

6. Wallace, R. L. Jr.: "The Reproduction of Magnetically Recorded Signals," The Bell System Technical Journal, Oct. 1951, pp. 1145-1173.

7. Eldridge, D. F. and E. D. Daniel: "New Approaches to AC-Biased Magnetic Recording," IRE Trans. on Audio, Vol. AU-10, May-June 1962, pp. 72-78.

8. Eldridge, D. F.: "Magnetic Recording and Reproduction of Pulses," IRE Trans. on Audio, Vol. AU-8, March-April 1960.

9. Kornei, Otto: "Structure and Performance of Magnetic Transducer Heads," Journal of the Audio Engineering Society, Vol. 1, No. 3, July 1953.

10. Eldridge, D. F.: "Magnetic Tape Production and Coating Techniques," Memorex Monograph No. 4, 1965, Memorex Corporation.

Chapter 5

Amplifiers & Equalization

The record amplifier, bias oscillator and playback amplifier do not differ very much from conventional circuitry design, except that particular attention must be paid to such factors as distortion, noise, and equalization. The block diagram in Fig. 5-1 illustrates how input voltage Vin is transferred into a current and passed through a pre-equalization network into the head winding. The high-frequency bias signal is added through capacitor C1. The bias signal may cause intermodulation distortion at the output stage of the record amplifier; consequently, a bias trap L2-C2 is inserted between the amplifier and the recording head.

During playback, the flux from the recorded tape is differentiated in the reproduce head, amplified and then equalized. The quality of a recording can be monitored by a metering circuit, which can be switched to the playback position for an A-B comparison between recorder input and output. The level indicator also can be used for monitoring the record and bias current through the resistor (typically 10 ohms) in the ground leg of the record head–a useful feature during checkout or service of a recorder. The lower portion of Fig. 5-1 is a level diagram for the signal as it passes through the record amplifier, is recorded, played back, equalized, and amplified.

RECORD AMPLIFIER

The record amplifier is an amplifier with an essentially flat frequency response that converts the input voltage into a record current. Its input impedance varies from 91 ohms for corresponding sensitivities range from 1 volt RMS to 1 millivolt RMS, which in either case gives an input sensitivity on the order of 10 milliwatts. The record amplifier should be designed to provide a current that will saturate the tape prior to amplifier distortion. The record currents listed in Fig. 4-19 are typical for a 1 percent harmonic distortion level in the tape and, as a rule of thumb, the saturation level is four times that level (12db).

The **record head** is essentially an inductive load for the record amplifier and its impedance, consequently, increases with frequency. This in turn requires that the record head be driven by a constant-current source which again dictates that the record amplifier has a high output impedance. A constant record current should, in reality, mean a constant record flux but eddy-current and hysteresis losses make it necessary to increase the record current toward the higher frequencies in order to obtain a more constant flux. The effect of eddy-current losses is illustrated in Fig. 5-2, showing how the head inductance decreases with increasing frequency. The reduced efficiency of the recording head means that equalization is required.

PRE-EQUALIZATION

Pre-equalization in the record amplifier is necessary not only because of the losses in the record head but also by the

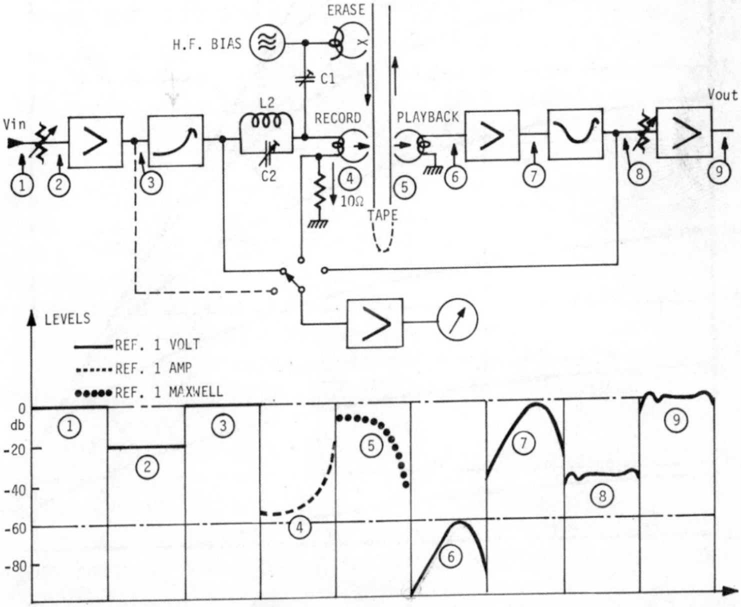

Fig. 5-1. Block diagram of the electronics in a typical recorder, and a signal level diagram. (Each section shows level vs frequency.)

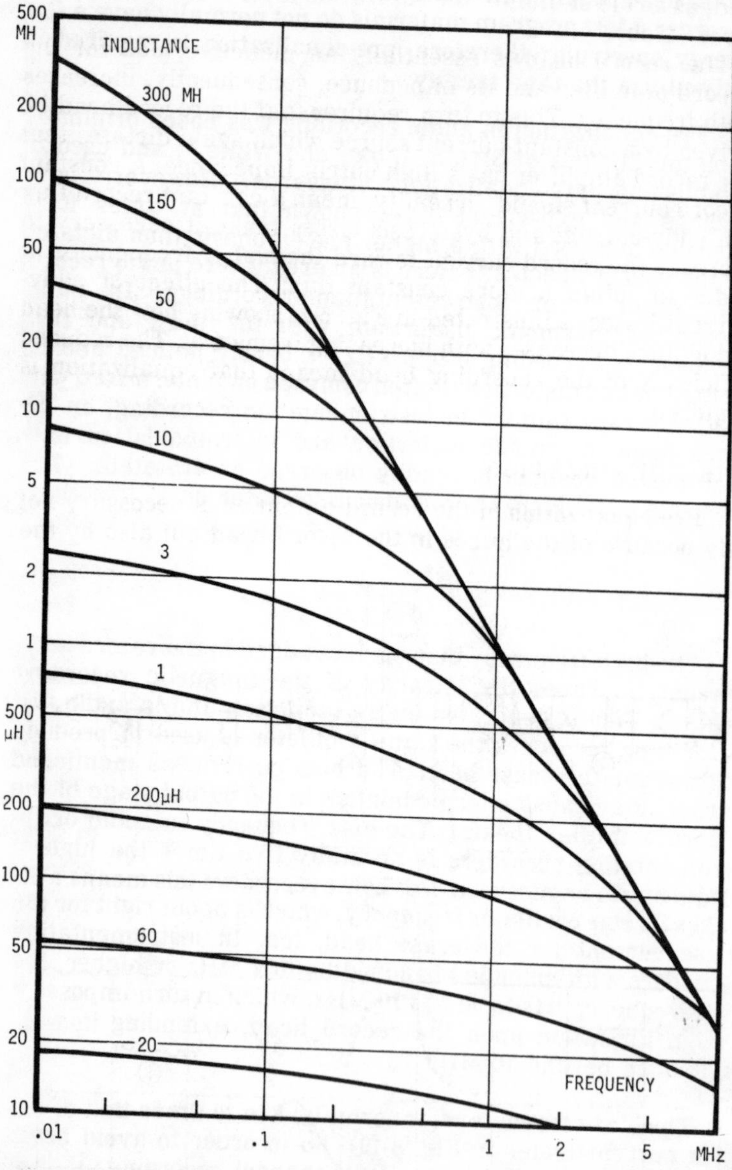

Fig. 5-2. Curves showing typical variations in head inductances vs frequency. The values listed in each case are inductances at 1 kHz.

nature of the program material. The losses in a laminated mu-metal head requires an 8 to 10 db boost at 2 MHz in a wideband recorder. Most program materials do not normally have a flat energy spectrum; therefore, pre-equalization is required to fully utilize the tape facility.

Pre-equalization in audio recording was based originally upon the findings of Sivian, Dunn, and White [1], and used in establishing pre-equalization standards. However, later findings and experience have shown that at times it is tolerable to record with a higher pre-emphasis than dictated by the energy spectra [2]. Here is an area where audio recordings differ from instrumentation recordings. In audio recordings, harmonic distortion, with the third and fifth components exceeding 2 to 3 kHz, is barely noticed and intermodulation distortion is often confused and intermixed with that of scrape flutter. In instrumentation recording, on the other hand, harmonic distortion and intermodulation in a multiplexed channel is readily observed as crosstalk. (The tolerable pre-equalization is discussed later on in this chapter.)

BIAS OSCILLATOR

The high-frequency bias current added to the record-head current improves the linearity of the magnetic recording process. Bias is generated by an oscillator, and in audio and home-type recorders the same oscillator is used to produce current for the erase head. The bias current, as mentioned earlier, may cause intermodulation in the output stage of the record amplifier (beats). The bias frequency in audio or instrumentation recorders is normally five times the highest frequency to be recorded. For home recorders this means a 60- to 75-kHz bias oscillator frequency, which is about right for the erase current for the erase head, too. In instrumentation recorders with extended bandwidths to 2 MHz or higher, the bias frequency has to be 7 to 10 MHz, which in turn imposes a design limitation upon the record head, extending its self-resonance beyond 10 MHz.

The bias current level is normally 8 to 10 times that of the data current (refer to Fig. 4-19). So in order to avoid beats between bias oscillators in multi-channel recording, it is a common design practice to use one master oscillator with buffer amplifiers in each record-amplifier section. This applies in particular to instrumentation recorders where as many as 14 channels may be recorded simultaneously.

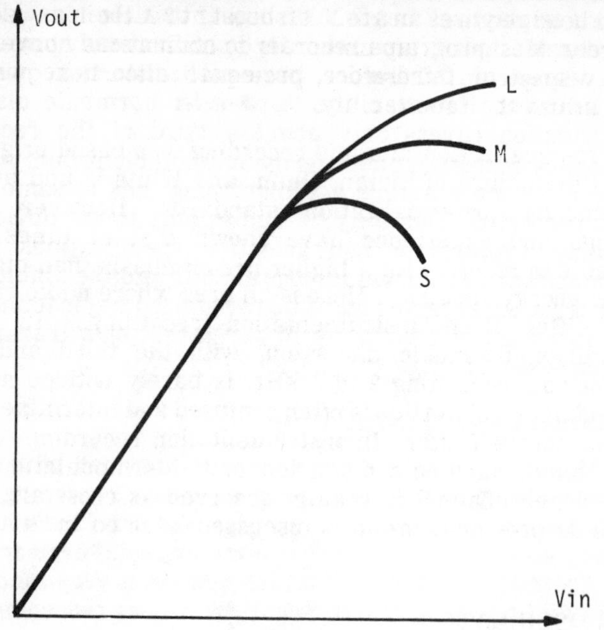

Fig. 5-3. Transfer characteristic curve for magnetic recording. L, long wavelengths; M, middle; and S, short wavelengths.

While high-frequency bias added to the record current improves the linearity of analog magnetic recordings, its waveform must be purely sinusoidal in order to avoid excessive noise[3] and distortion.

DISTORTION

Like any component in a transmission system, magnetic tape overloads with excessive signal levels. This causes **harmonic** and **intermodulation distortion**, which can be related to the overload curve for that particular tape. Lack of symmetry in the bias current waveform may likewise cause distortion. Fig. 5-3 illustrates overload curves for three different frequencies (which in turn can be related to recorded wavelengths).

The transfer function curves in Fig. 5-3 give rise to uneven-order harmonics and intermodulation distortion, and an asymmetrical bias current waveform causes even-order harmonic distortion. (This may also be caused by magnetized record or playback heads.) Assuming that the bias oscillator waveform is pure, tape recorder specifications normally list the degree of third-order harmonic distortion, which is predominant. However, the third-order harmonic distortion specification covers only about a third of the recorder's frequency response, and intermodulation distortion, which is quite often more severe and disturbing, has to some extent been neglected. This may be due in part to the difficulty in defining appropriate and sensitive specifications for intermodulation distortion measurements and the ambiguity of such tests. An accepted but little recognized test is found in the measurement of the "noise power ratio" of a transmission system (see Chapter 10).

RECORD LEVEL INDICATOR

Magnetic recorders are generally provided with a **level indicator** to warn the operator of excessive recording levels. In its infancy, the magnetic recorder was provided with a VU meter, which stands for volume meter, a leftover from the early broadcast industry. This instrument is very inadequate for informing the engineer about the proper recording level, since distortion may very well take place at instantaneous peaks of the signal to be recorded, in turn, causing intermodulation distortion. Peaks are much more readily detected by peak-reading indicators, which may be a moving-coil instrument with a suitable amplifier, a "magic" eye instrument, or neon lamp.

A record-level indicator should be connected (as shown in Fig. 5-1), to the electronic circuitry after equalization, since constant-current recording (which produces a constant flux) provides a nominally constant distortion level. The disadvantage of connecting the level indicator at this point is that it will not show a true comparison between Vin and Vout (A-B test). A fourth position, as shown by the broken line in Fig. 5-1, would overcome this.

PLAYBACK AMPLIFIER

The **playback amplifier's** function is to amplify the weak reproduce head voltage, equalize the signal, and further amplify it to a suitable output level. The stages in the reproduce amplifier are normally of conventional design, with the exception that extreme care must be taken in the selection of the input transistor immediately following the reproduce head. The available voltage from the reproduce head is indicated in Fig. 4-19, which at extreme low frequencies (50 Hz) is in the microvolt region. This means that the input stage in the playback amplifier must have an equivalent internal noise voltage that is only a fraction of a microvolt. Since the reproduce head is an inductive generator, it is further important that the input impedance be high to avoid roll-off at high frequencies.

The output voltage from the reproduce head can be increased by raising the number of turns on the reproduce head core. But this lowers the self-resonance of the head and limits the high-frequency response. Therefore, the design criteria becomes a compromise between bandwidth and output level, a compromise is illustrated in Fig. 4-18 and now we have the additional restriction of the limited number of turns that can be wound on the head core. The self-resonance of a magnetic head can readily be measured, as outlined in Chapter 10. In a practical design, care must be taken to include the head cable and amplifier input impedance. As in any inductive element, the reproduce head has a certain quality, called Q. The Q may be somewhere around 5 to 10 at low frequencies, decreasing toward higher frequencies due to eddy-current losses in the lamination, and can be as low as one at frequencies of 2 MHz. In some instances, the low Q at high frequencies can be advantageous if the head resonance is "tuned" to the upper limit of the desired frequency response. However, in pulse work this may be a disadvantage since the resonant frequency introduces an additional phase shift in the playback signal and it may be desirable to have the playback amplifier input impedance designed to a value that the combination head inductance and input capacitance are critically altered. The proper value of the input impedance can easily be determined with the test setup shown in Fig. 5-4. A square-wave generator is connected to a 10-ohm resistor in the ground leg of the reproduce head. (Assuming that the normal load for the square-wave generator is 600 ohms, a 590-ohm resistor must be used in series with the connection.) With a one millivolt peak-to-peak signal across the 10-ohm resistor, the variable input resistor (R) is adjusted until the amplified square wave shows little or no ringing.

Another factor to be considered in the design of the first stage in the playback amplifier is the large dynamic range it is required to handle. A high signal-to-noise ratio is particularly difficult to achieve at the long wavelengths. At the same time, the first amplifier stage must be able to handle signals 40 to 50 db higher than the low frequencies.

POST-EQUALIZATION

The playback head output voltage, as illustrated in Fig. 4-10 and 4-19, requires equalization in order to obtain a flat amplitude versus frequency response. An amplitude-versus-frequency-correcting network, with an inverse frequency response compared to the head output, may be inserted as a passive loss element between two amplifier stages or it may be incorporated in a feed-back loop. It is important to notice that several of the losses in the record-playback process are not associated with a phase shift, while any passive network made up of R, L, and C components will exhibit a phase shift. For example, from electrical network theory it is well known that an RC network has a 6 db per octave slope associated with a 90-degree phase shift. Therefore, the equalization network introduces a phase shift which is detrimental to a true reproduction of the initial data.

This is illustrated in Fig. 5-5 where the input waveform is assumed to be a square wave. A square wave is essentially expressed as:

$$V = v1 \cos \omega 1\ t + \frac{V1}{3} \cos \omega 3\ t + \frac{V1}{5} \cos \omega 5\ t + \text{-----}$$

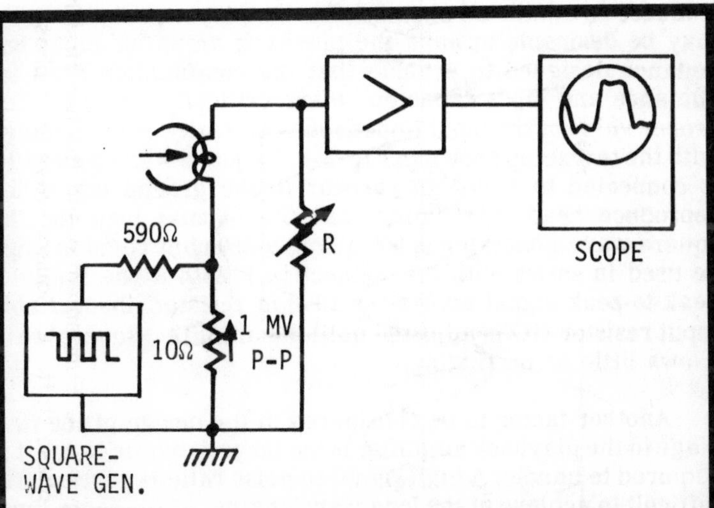

Fig. 5-4. Test setup for determining input impedance.

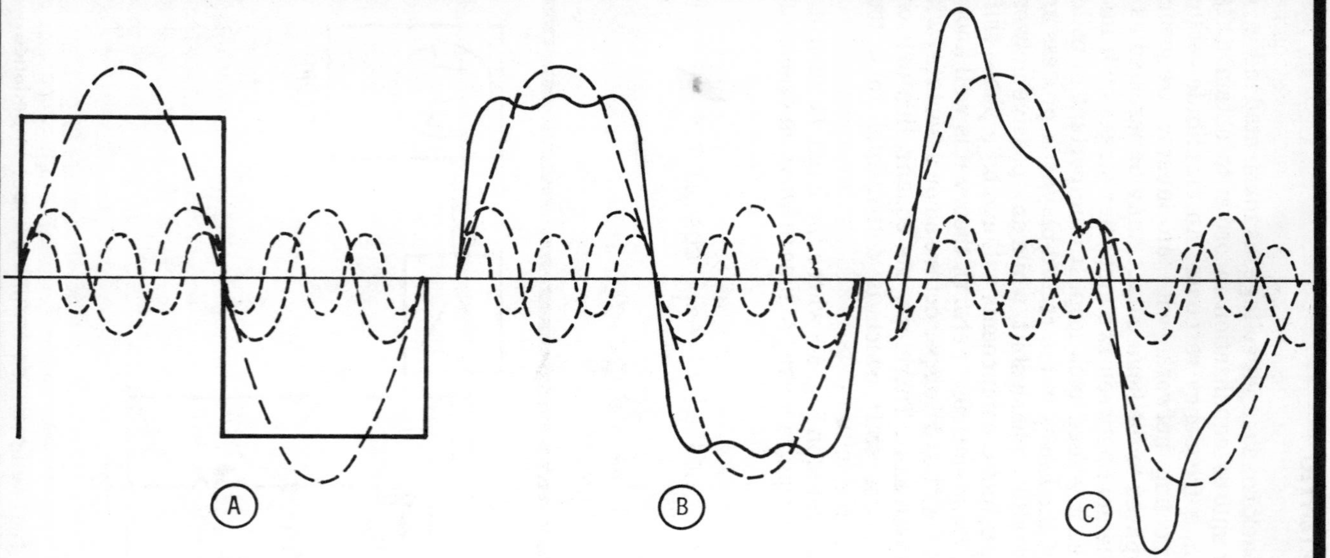

Fig. 5-5. A. Fundamental and the two first harmonics in a square wave. B. Reproduced waveform with no phase shift. C. Square wave with phase shift.

For simplicity, in Fig. 5-5 we are considering only the fundamental, third, and fifth harmonics in composing the square wave. This is shown in Fig. 5-5B, where we have correct amplitude and phase relationship between the components. If the amplitude relationship remains correct, but phase shift is introduced, then the resultant is a distorted square wave, illustrated in Fig. 5-5C; an actual photograph from an oscillograph screen is shown in Fig. 5-6.

Accurate data recording and playback can only be accomplished using a **post-equalization** network that does not introduce phase shift. Such networks are normally of a delay-line type, an all-pass lattice network, or the type shown in Fig. 5-7. The essential element is a 1:1 transformer and the network performs the following: 1) Low-frequency input voltages are integrated across C. 2) This function is shelved at Fm by R1, and toward higher frequencies the signal is now differentiated across L1, with a subsequent increase in amplitude. 3) The phase has simultaneously been twisted 180 degrees but is reversed by the opposite coupling of L1 and L2. 4) The additional boost at band edge is accomplished by tuning L2 with C2, damped by R2. A variable value of R1 makes it possible to shift midrange equalization, C2 determines the band-edge frequency, and R2 the amount of boost.

In addition to the equalizer phase shift, there is, literally, also a phase shift in the tape flux itself. This was mentioned in Chapter 4 and can be proved by recording a square wave on a tape, controlling the amplitude and then equalizing the phase shift of the signal. When the tape is played back in the reverse direction the distortion of the waveform is readily seen. The waveform becomes highly distorted, which explains the degradation ocurring in tape copies made in sequences, where the transient response (like in piano music) of the sixth copy, for example, has been completely distorted.

EQUALIZATION STANDARDS

As established previously, equalization is needed partly to overcome record head losses and also to flatten the reproduce head response. This equalization scheme is known as **constant-flux recording**; however, it must be kept in mind that the external flux from the tape does decrease with decreasing wavelengths in accordance with the thickness loss formula. Constant-flux recording is used in virtually all instrumentation recorders, where quite often frequency-multiplexed signals are recorded. The nominal levels of the channels are identical, as shown in Fig. 5-8, and any pre-

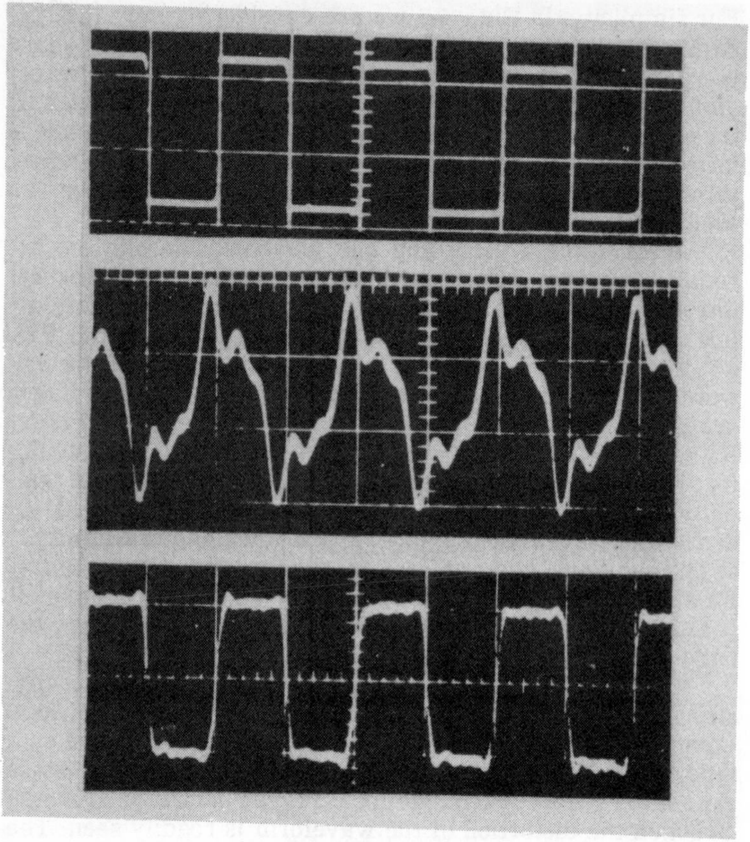
Fig. 5-6. Top: Square-wave input. Center: Typical playback wave form. Bottom: Properly phased equalized playback.

equalization of the recording would result in overload and intermodulation distortion.

In audio recording the situation is quite different, since the energy spectrum of music falls off quite rapidly after a peak of around 500 Hz. Since post-equalization enhances low-frequency amplifier noise, high-frequency tape hiss, and head noise, it was early recognized that pre-equalization would allow for better utilization of the tape and simultaneously reduce the necessary post-equalization and associated noise. The allowable pre-equalization depends on the type of music to be recorded since different instrument groupings and musical compositions have different energy spectra. In order to have interchangeable tape recordings the need for standards is obvious. There is a general agreement that any standard must

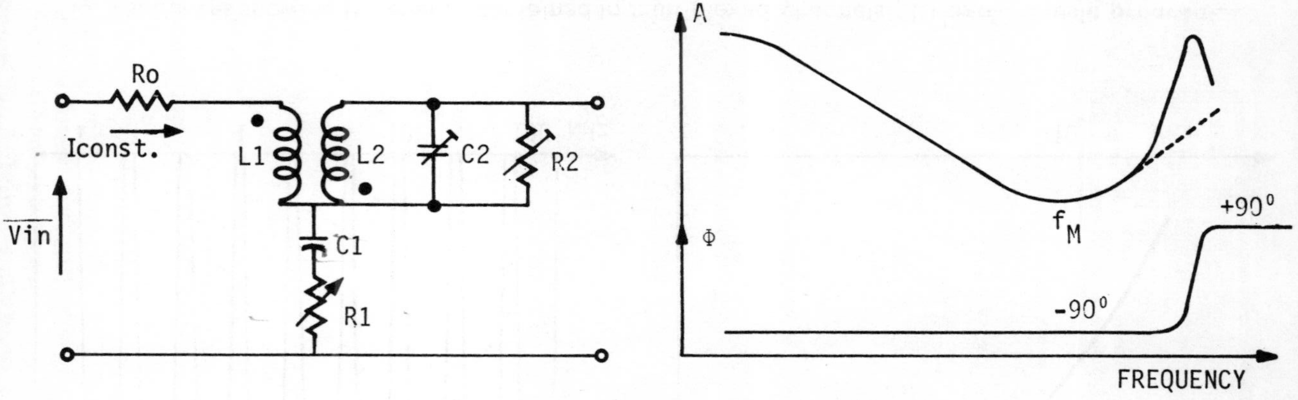

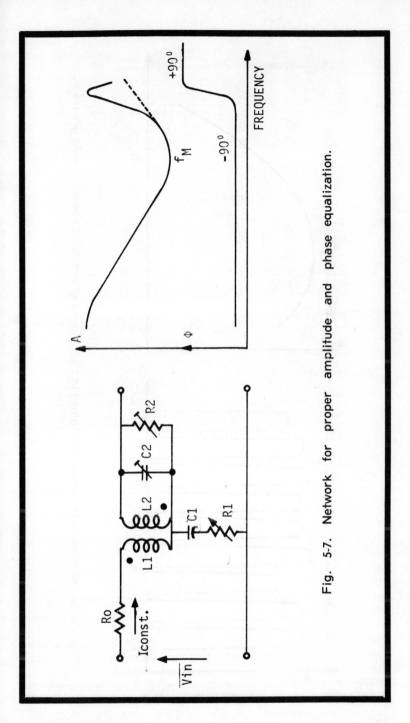

Fig. 5-7. Network for proper amplitude and phase equalization.

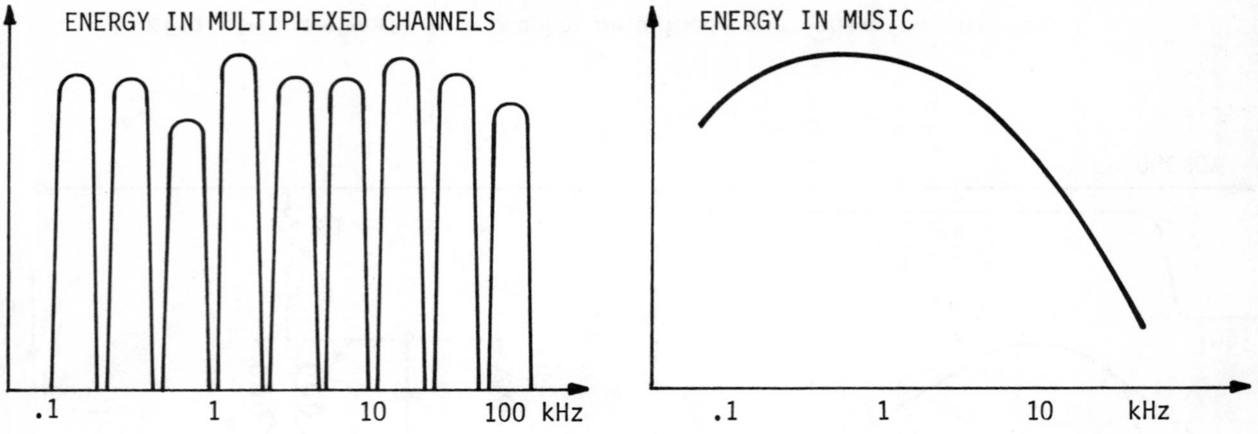

Fig. 5-8. Curves showing the energy contained in multiplexed channels (L) and a music program.

pertain to the remanent flux on the recorded tape and it is common among the standards that there is an adherence to a flux level versus frequency response that essentially follows the coating thickness law. This law is not frequency but wavelength dependent, which in turn means that equalization networks (or component values) must be changed with different tape speeds. The variance in standards is found in the selected cross-over frequency values for the tape flux.

Fig. 5-9 illustrates three widely used standards at the tape speed of 7½ IPS with a spread in cross-over frequency of up to 2:1, which gives a 6 db level difference at the high frequencies. A few recorders offer a selector switch between, for example, CCIR and NAB, although this may be somewhat academic since most recordings are tailored to the individual recording engineer's taste, which includes his control of bass and treble. And the individual choice of the standardizing organizations is also evident in the fact that NAB in the U.S.A. recommends a low-frequency boost during recording with the argument that less post-equalization is required with less amplifier hum and low-frequency noise, therefore, a lower general noise level. On the other hand, CCIR in Europe argues that this very easily results in over-recording and intermodulation in program materials that are rich in low-frequency energy, such as organ music.

Improved tapes may result in CCIR changing to the IEC curve, but the different standards are still nevertheless inconsistent: different concepts are used and the same terms have conflicting uses, resulting in confusion when standards are compared. (See Reference 5 listed at the end of the Chapter for an excellent discussion of standards.) Chapter 10 contains a listing of the various standards at tape speeds of 30, 15, 7½ and 3¾ IPS.

Post-equalization in a tape recorder is most easy to adjust by the use of a standard alignment tape, which contains a high-frequency tone for playback head azimuth alignment, a normal record level tone (0 db), and a series of different frequency tones for equalizer adjustment. The next step in adjusting the electronics is setting the bias level for the selected tape, and finally the trimming of the pre-equalization network to obtain an overall flat level versus frequency response. These steps are covered in detail in Chapter 10.

NOISE

Several noise sources exist with the magnetic record/playback process, originating in the electronics as well as emanating from the tape itself. Ideally, one would like the

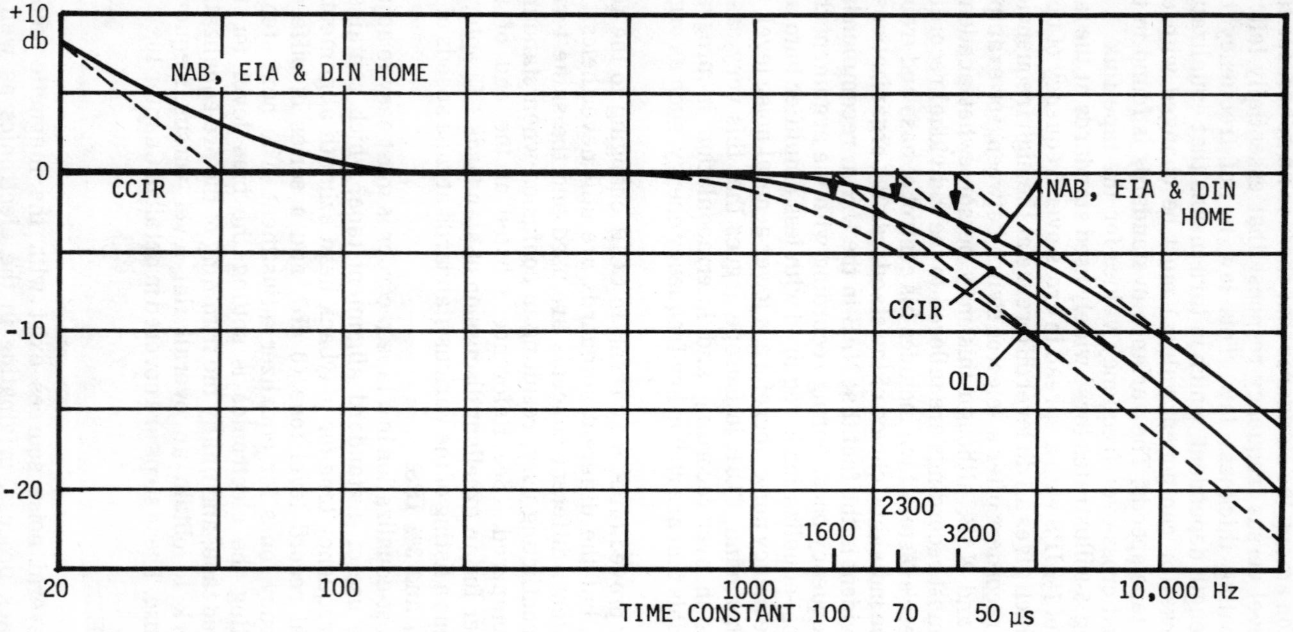

Fig. 5-9. Magnetic tape flux standards at 7½ IPS.

recorder to be absolutely quiet with the tape as the only noise source and a quality recorder today is the result of step-wise improvements of the transport, heads, electronics, and tape.

Electronics noise is quite trivial, originating from power supply hum, low-frequency noise in the input stage, and resistor noise (hiss). These noise generators can be fairly well controlled by careful design and component selection, such as low-noise resistors and high-grade transistors and capacitors. If, for example, an insertion-type equalizer is used after the preamplifier, great care must be taken in optimizing the signal level and equalizer impedance in order to avoid additional noise from the resistive elements in the equalizer.

High-frequency noise in a wideband recorder (1 to 2 MHz) is primarily generated in the reproduce head itself, caused by eddy-current loss, and the equivalent loss resistance of a playback head is typically on the order of 1K at 1 MHz, which is higher than the equivalent input noise resistance of a well designed amplifier.

Electronic noise sources are aggravated by the required post-equalization boost of low as well as high frequencies. And here a word of caution when evaluating a recorder's **signal-to-noise ratio**: it is normally measured without filters or a weighting network. The measured noise voltage is representative for the low- and high-frequency noise and does not show that the mid-band noise is much lower, in the order of 20 to 30 db for an audio recorder and as much as 50 db for a constant-flux instrumentation recorder. It would be entirely proper to use a weighting network when evaluating an audio recorder, since the shape of the noise spectrum happens to be close to Fletcher-Munson's curves for the ears' sensitivity. The subjective signal-to-noise ratio, therefore, is closer to 70 to 80 db for a recorder that measures 50 db on a VTVM. It is likewise more useful to evaluate an instrumentation recorder from a spectrum analysis of its noise, rather than by a single broadband measurement.

The other major noise source in a tape recorder is the tape itself and it shows up in various ways that are characteristic for magnetic recording. If a **virgin tape** is played on an audio recorder, one will notice a hissing sound. This noise is caused by the limited and varying number of particles that the playback gap "sees." Each particle is a minute permanent magnet and if a sufficiently large and constant number were "seen" by the playback head, their fields would cancel out and no noise voltage generated.

If the same tape again is played on the recorder, this time with the record button pushed, the noise level will be several db higher. This **bias-induced** noise remains a major limitation

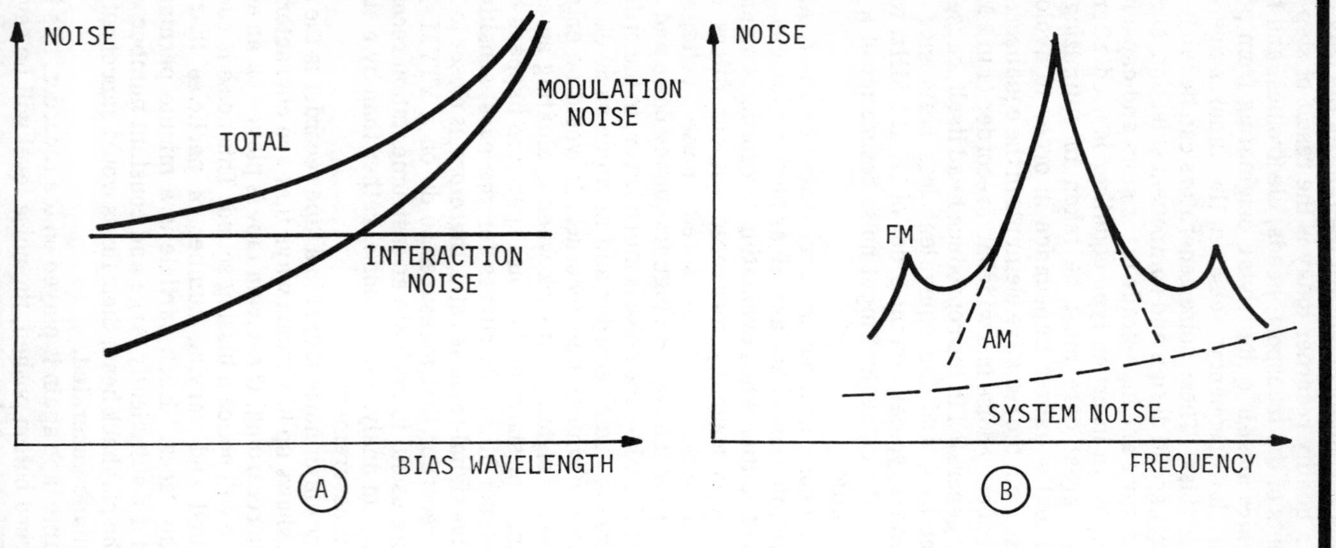

Fig. 5-10. A. Bias noise as a function of bias wavelength. B. AM vs FM modulation noise.

of the signal-to-noise ratio in modern recorders and can at best be reduced to a level of 3 db higher than virgin tape (or bulk-erased tape)[6]. Investigations of this noise indicate two categories of bias noise.

When the bias frequency is higher than that resulting in a wavelength of 80 microinches, noise appears to arise from a combination of the bias field and the interaction field between particles. A lower bias frequency does not only increase the noise level, but is now generated as AM-**modulation noise.** This is illustrated in Fig. 5-10A. AM-modulation noise (or noise behind the signal) is best understood if for a moment we consider the DC-magnetization noise of a magnetic tape. A DC-magnetized tape can have a noise level 20 to 30 db higher than a virgin tape. This is due to non-uniformity of the coating dispersion, the backing surface, and the coating surface. An AC signal likewise generates noise and can be considered a slowly varying DC signal. The noise is formed as sidebands around the signal frequency, as shown in Fig. 5-10B. This noise is not very noticeable, since it is masked by the signal;only when pure tones are recorded and played back can the ear detect AM noise.

Returning to the bias-induced noise, it is likely that a low-frequency bias generates AM noise with sidebands reaching into the pass band of the recorder. These findings were made with a distortion-free bias oscillator[6]. Additional noise will be generated if the bias signal contains even-harmonic components, which in essence constitute a DC signal for which we have already explained the noise mechanism. The sound of this noise is an intermittent, gurgling-type–quite different from hiss. It is best avoided by using a push-pull type oscillator, possibly with filtering. Some recorders use a bugging DC current through the record head winding, but this is only a cure, not a solution. AM-modulation noise is often mistaken for FM-modulation noise, which is caused by scrape flutter in the transport mechanism. This generates sidebands as shown in Fig. 5-11 and can be very disturbing in changing the sound of a pure tone into a harsh sounding note.

EFFECTS OF AMPLITUDE VARIATIONS

Direct analog magnetic recording is limited in performance by tape coating variations (non-uniform dispersion, limited number of particles "under" the reproduce gap, and coating thickness variations) which cause **amplitude variations** that in turn limit the accuracy of the recorder/reproducer. A representative figure for amplitude variations is plus or minus .5 to plus or minus 1 db.

A figure of merit for a transmission system is the product of bandwidth and information capacity, as expressed in Shannons Law for the channel capacity:

$$J = B \times \log_2 (1 + Ps / Pn)$$

$$= 2B \times \tfrac{1}{2} \log_2 (1 + Ps / Pn)$$

where B equals the system bandwidth, Jo equals ½ log2 (1 plus Ps/Pn) equals the maximum possible quantity of information Ps/Pn equals signal-to-noise ratio. This fundamental law assumes a Gaussian distribution of the noise and absence of amplitude variations and, therefore, cannot be applied meaningfully to tape recorders without consideration of its limitations.

The noise spectrum for a tape recorder shows the poorest signal-to-noise ratio at the band edges in the 20 to 35 db range and the best at mid-frequencies in the range of 75 to 95 db for a 200-Hz bandwidth. Therefore, the actual channel capacity will be better than the one calculated when using the specified overall signal-to-noise ratio, which typically is in the 20 to 30 db RMS-to-RMS range for wideband recorders (1.2 to 2 MHz).

Fluctuations in the tape-coating characteristics present a limitation of the number of amplitude levels that can be distinguished from each other. If the amplitude modulation is plus or minus M (the ratio of amplitude variation MVN/VN), the difference between two separate amplitude levels must be at least 2M. For two distinguishable voltage levels, we must have:

$$Vn\ (1-M) - V = \frac{\text{noise peak}}{Vn-1(1+M)+V\text{noise peak}}$$

which results in an amplitude difference of:

$$\Delta vn = 2MV + 2V\text{noise peak}$$

When:

$$2MVn \gg \frac{\text{noise peak}}{2V\text{noise}}$$

which normally is the case for a large SNR, then:

$$\Delta V \cong SMVn$$

and the maximum quantity of information is then:

$$.Jo = \log_2 (SNR/2M)$$

The formula for Jo should be used when the SNR is greater than one-half M while Shannon's Law still gives the limitations for lower SNRs.

The quantity of information has been calculated for four values of M and is shown in Fig. 5-11. At low SNRs Shannon's Law applies, but at higher values the recorder performance is limited by amplitude variations. The right-hand scale shows the equivalent instrument accuracy. A typical performance specification for a 1.5-MHz recorder/reproducer is 25 db SNR, with M equalling plus or minus 1 db and this corresponds to an instrument accuracy of about plus or minus 4 percent. Improvements in SNR will give a diminishing return unless means are found to also reduce the amplitude variations.

The amplitude variations are mainly a result of coating

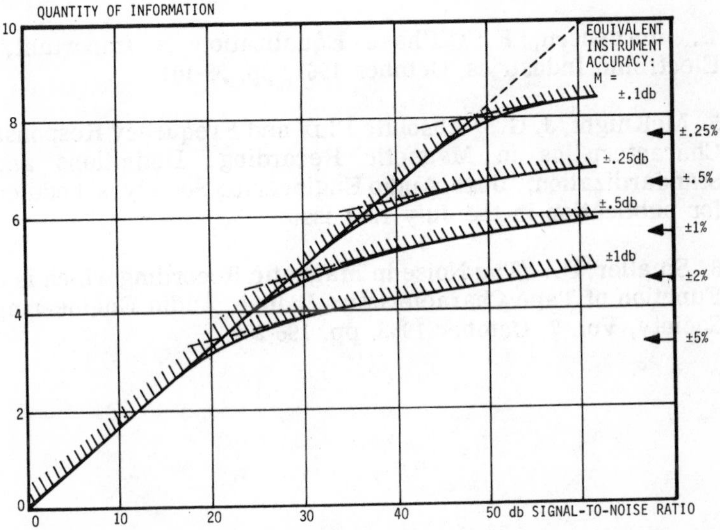

Fig. 5-11. Quantity-of-information limits imposed by Shannon's Law as a function of a system's signal-to-noise ratio. Amplitude modulation further restricts this quantity and the equivalent instrument accuracy.

thickness variations and tape surface conditions, the latter causing varying spacing losses during recording and reproduction. Improvements in controlling the tape coating are being pursued by the tape manufacturers while new type magnetic heads with a better controlled field for both recording and reproduction are being sought by the recorder/reproducer manufacturers and an increase to plus or minus 2 percent accuracy is becoming a practical reality.

References:

1. Sivian, L. J., H. K. Dunn, and S. D. White: "Absolute Amplitudes and Spectra of Certain Musical Instruments and Orchestras," Jour. Acoustical Society of America, January 1931, p. 330.

2. McKnight, J. G.: "Signal-to-Noise Problems and A New Equalization for Magnetic Recording of Music," Jour. Audio Engineering Society, Vol. 7, January 1959, pp. 5-12.

3. Ragle, H. U. and P. Smaller: "An Investigation of High-Frequency Bias-Induced Tape Noise," IEEE Trans. on Magnetics, June 1965, pp. 105-110.

4. Jorgensen, F.: "Phase Equalization is Important," Electronic Industries, October 1961, pp. 98-101.

5. McKnight, J. G.: "Absolute Flux and Frequency Response Characteristics in Magnetic Recording: Definitions and Standardization," Jour. Audio Engineering Society, scheduled for publication in the July 1967 issue.

6. Smaller, P.: "The Noise in Magnetic Recording which is a Function of Tape Characteristics," Jour. Audio Engineering Society, Vol. 7, October 1959, pp. 196-202.

Chapter 6

Selection of Tapes & Accessories

Magnetic recording tapes can basically be divided into four groups: audio, computer, video, and instrumentation tapes. Within each group there are again several varieties, primarily different in playing time, price, or special applications. These differences may be confusing at first, so let's consider the basic requirements in each case and the subtle characteristics of each group.

The requirements for an ideal magnetic tape are as follows: The coating shall be completely uniform and have a perfectly flat surface, which assures good contact with the recording and playback heads. The combination of the coating and film base shall be completely pliable and at the same time possess adequate mechanical strength, so stretching or breaking of the tape is prevented. And finally, it should be completely insensitive to storage, temperature changes, and humidity changes.

PROPERTIES OF BASE MATERIALS

All magnetic tapes on the market today use one of the following three materials as a base: cellulose acetate, polyvinyl chloride, and polyester. **Cellulose acetate** film (acetate) is a popular base for audio tapes. But for the demanding use in precision magnetic tape, it lacks the high tensile strength, dimensional stability, and tear resistance. **Polyvinyl chloride** (PVC) is used to a large extent in Europe as a base for audio recording tape. Its tensile strength is slightly lower than that of cellulose acetate. **Polyester** film is between two to three times stronger than any of the two other materials and it is twelve times more stable dimensionally, three times more resistant to tear and about ten times more resistant to cupping, curling and humidity absorption.

Cupping is illustrated in Fig. 6-1. Two tape samples are placed on a plane supporting surface; the cupped tape will

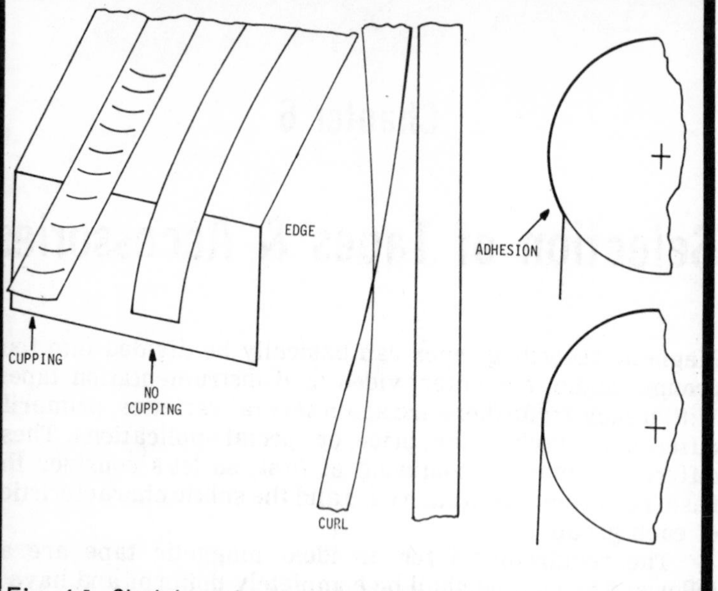

Fig. 6-1. Sketches showing the effect of cupping, curl, and layer-to-layer adhesion.

stand straight out, while the better tape will tilt in a smooth arc. Cupping is generally found in inferior tape and is due to improper manufacturing or differences between coefficients of thermal or hygroscopic expansions of coating and base film. Curling is also illustrated and results in a twist of a free hanging length of tape. This is again due to poor manufacturing where differential stresses can build up in the coated web; it is aggravated by a bad slitting process.

The difference between polyester (also called Mylar) and acetate is illustrated in Fig. 6-2. The left portion shows a stress-strain diagram clearly indicating the higher strength of the polyester. It also shows that when the acetate breaks, there is a small elongation, and this has a practical application since the tape can be readily spliced together with little distortion in the recorded material. The polyester tape, however, will elongate considerably (100 percent) before it breaks and this, of course, makes the stretched portion useless. The Table in Fig. 6-2 shows some mechanical properties of the various base materials for a 1 mil thick one-fourth-inch tape.

Although polyester base materials are twice as expensive as any of the other two, it is employed exclusively in the

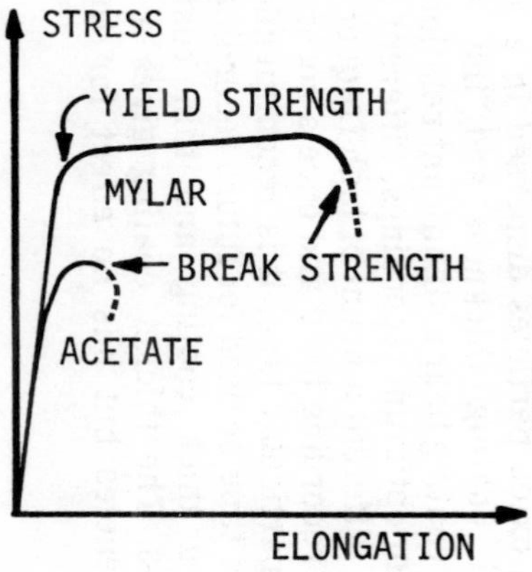

		ACETATE	PVC	MYLAR	TENSILIZED MYLAR
YEILD STRENGTH	LBS.	2.2	1.8	3.5	5.5
BREAK STRENGTH	LBS.	2.5	2.0	6.5	10.0
COEFF. OF THERMAL EXPANSION	IN/IN/°F	$2x10^{-4}$	-	$1.5x10^{-5}$	$3.3x10^{-5}$
COEFF. OF HYGROSCOPIC EXPANSION	IN/IN/%RH	$2x10^{-4}$	-	$1.1x10^{-5}$	$2.2x10^{-5}$

Fig. 6-2. Mechanical properties of various base materials for a 1-mil thick ¼″ wide tape.

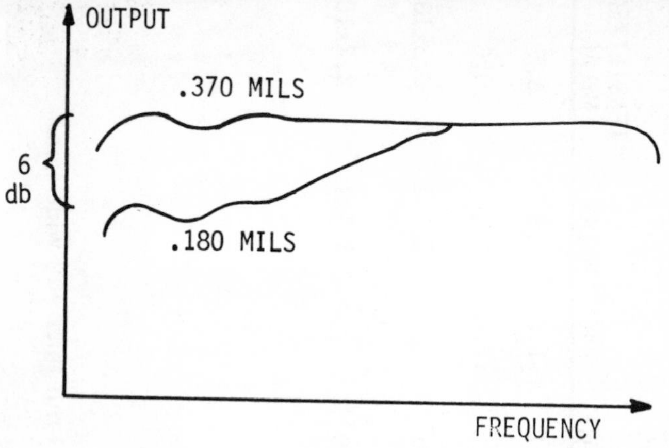

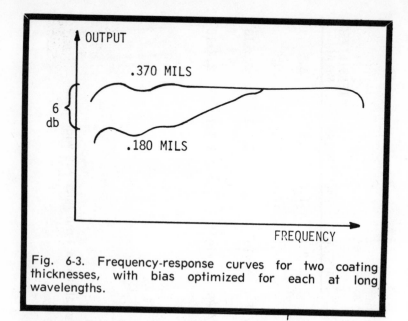

Fig. 6-3. Frequency-response curves for two coating thicknesses, with bias optimized for each at long wavelengths.

manufacture of precision magnetic tape for computers, video, and instrumentation. The strength of the polyester film can be increased further by pre-stretching it. The stretching orients the long chain molecules in the film in the direction of the stretch. Pre-stretched tapes are referred to as **tensilized** or tempered. The disadvantage of tensilized polyester films are that they have a memory and if they are heated beyond a certain point, the tape will shrink back close to its original size with consequent distortion of the recorded material.

PROPERTIES OF COATINGS

The magnetic coating on the tape is almost exclusively fabricated from iron oxide particles dispersed in a suitable binder. The overall **coating thickness** and the surface characteristics of the coating bear a significant relationship to the final recording. For optimum recordings, different coating thicknesses require a different amount of high-frequency bias currents through the recording head and once a recorder has been tuned up for a particular tape, it is recommended that that particular type of tape be used for future recordings. The difference between a thick coating and thin coating is illustrated in Fig. 6-3. The thicker coating gives a higher output at low frequencies but has no effect for the high frequencies.

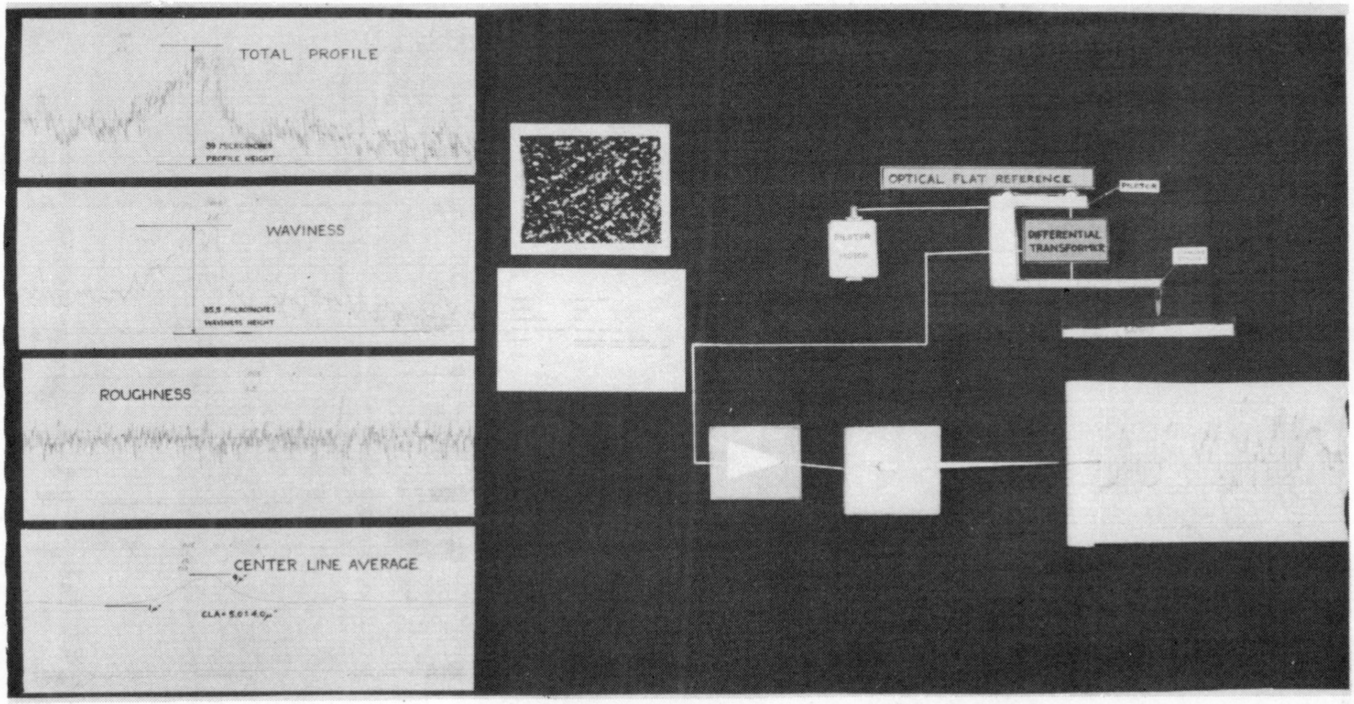

Fig. 6-4. Profile of a tape surface measured with a Proficorder. (Courtesy Memorex Corp.)

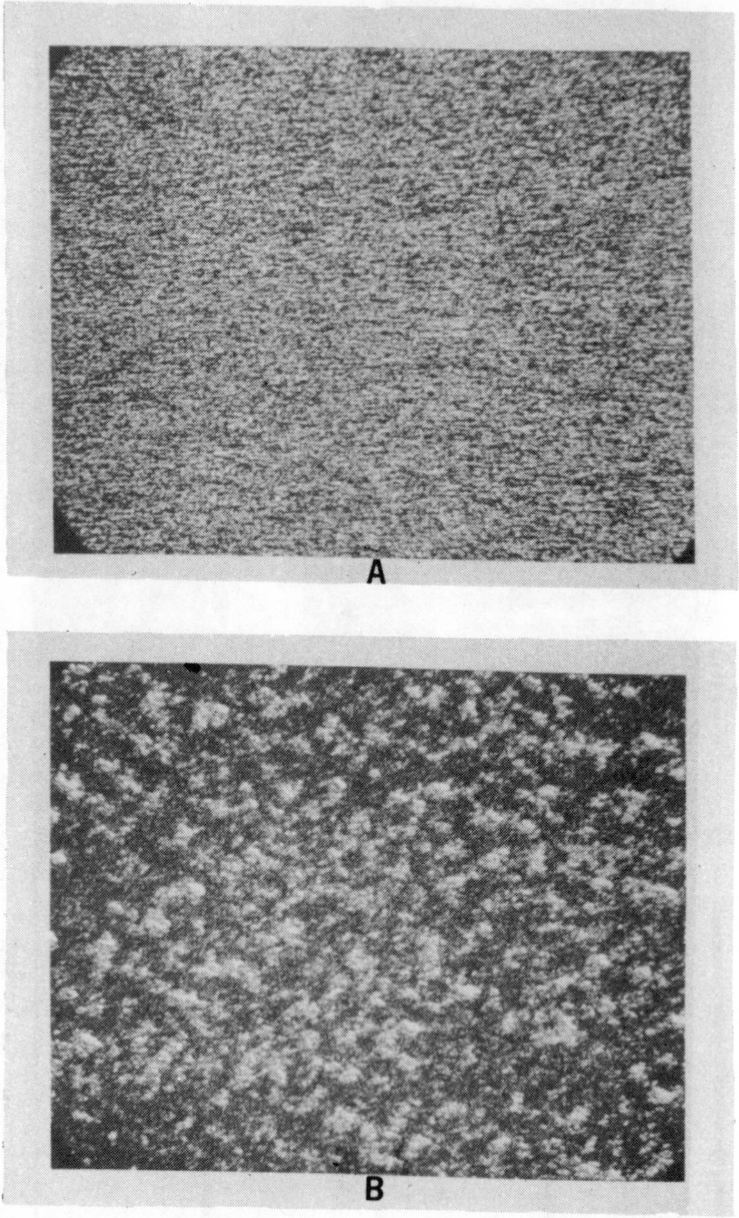

Fig. 6-5. Tape surfaces magnified 155 times. A, smooth; B, rough.

The surface characteristics have a significant bearing on the high-frequency performance of a magnetic tape. Any spacing between the magnetic heads and the tape surface causes a reduction of high-frequency (short wavelengths) response, and a rough tape surface, in effect, acts as a spacing between the magnetic coating and the tape. In the manufacture of magnetic tape it is a common practice to surface-treat the tape to obtain a finished product that has as smooth a surface as practical.

Surface flatness can be evaluated by means of a so-called Proficorder. This instrument places a very small diamond needle across the tape surface and any motion of the needle (like in a phonograph pickup) is amplified and the output waveform plotted on a paper chart. An example is shown in Fig. 6-4. Although it can be debated that the diamond stylus may deform the tape surface under the high unit area pressure, it is still a useful instrument in comparing various tapes. A magnified photograph of a typical tape surface is shown in Fig. 6-5, which shows indentations and protrusions in the tape surface.

Another important aspect of tape surface characteristics is its wear properties. A high degree of surface roughness causes excessive head wear in addition to the loss of high frequencies. The iron oxide used in the magnetic coating is in itself a highly abrasive material and an improperly manufactured and surface-treated tape may, in essence, be compared with sandpaper. This is often the case with the so-called "white box" tapes and the temptation to buy these because of their lower price should be resisted, since it eventually results in a much higher cost in replacing worn-out magnetic heads.

An ideally smooth tape does have an apparent disadvantage in causing an almost molecular adhesion to the head surfaces. This phenomena is chemical-mechanical in its nature. The effect demonstrates itself in two ways: gap smear or "varnish." Gap smear, which is particularly noticeable in microgap heads, will cause the core lamination material to flow across the gap and in essence short circuit it magnetically. The varnish phenomena is a microscopic buildup of a clear film on the magnetic heads which may cause separations between the tape surface and the head surface. This varnish may come from resins in the coating material and should be eliminated by other, non-varnish-forming components. A typical case of gap smear is shown in Fig. 6-6.

When used in video recordings, the tape must further exhibit good temperature characteristics. In a transverse-scan video recorder, the relative speed between the head and

Fig. 6-6. Gap smear effects.

the tape is in the order of 1500 IPS and the localized heat generated by friction can be very high and cause a breakdown of the coating. This is particularly pronounced in the helical-scan video recorders with still-frame capability; that is, the tape motion is stopped while the scanning head rotates and thereby produces a single-frame picture of the video information.

PROPER SELECTION AND TESTS

In choosing a magnetic tape, several factors should be considered:

Field of application
Response required
Playing time
Quality

The field of application will coarsely divide the tapes into four groups: audio, instrumentation, computer, and video.

Audio tapes are one-fourth-inch wide tapes used by the sound recording industry, broadcast stations, and in homes. (One-half and one-inch wide tapes are also used in certain sound recording studios, where several microphone outputs are recorded on individual tracks for later mixing.) The one-fourth-inch tapes are typically wound on 7-inch diameter reels or 10½-inch reels for professional use. The 7-inch reels contain 1200 feet (one and one-half mil base thickness), 1800 feet (1 mil), or 2400 feet (one-half mil tensilized polyester), and are generally called out as standard, long-play, and extra long-play (this call-out differs from manufacturer to manufacturer). The 1800-foot version seems to be the most popular, considering economy and playing time.

The quality of tapes on the audio market varies widely and it is generally recommended that the user buy brand-name tapes. The so-called white-box tapes may have been rejected by computer tape manufacturers; these tapes are likely to be high in abrasion and, in addition, quite likely to be under a strain resulting from a poor slitting process.

Variations in the frequency response of recognized tapes are minor, but may require different bias settings (and possibly equalization). It is, therefore, a good rule to stay with a given brand tape once it has been selected. There are several tests that can be made in the selection of a tape. These tests are visual and if a reel of tape does not pass them, it will most likely perform poorly on a tape transport.

SPEEDS IPS

LENGTH	HIGH SPEEDS		HIGH SPEEDS		MEDIUM SPEEDS		SLOW SPEEDS	
FEET	120	60	30	15	7-1/2	3-3/4	1-7/8	15/16
9600	16'	32'	1:4'	2:7'30"	4:15'	8:30'	17:	1*10:
7200	12'	24'	38'	1:36'	3:12'	6:24'	12:48'	1*1:36'
4800	8'	16'	32'	1:4'	2:7'30"	4:15'	8:30'	17:
3600	6'	12'	24'	48'	1:36'	3:12'	6:24'	12:48'
2400	4'	8'	16'	32'	1:4'	2:7'30"	4:15'	8:30'
1800	3'	6'	12'	24'	48'	1:36'	3:12'	6:24'
1200	2'	4'	8'	16'	32'	1:4'	2:7'30"	4:15'
900	1'30"	3'	6'	12'	24'	48'	1:36'	3:12'
600	1'	2'	4'	8'	16'	32'	1:4'	2:7'30"
450	45"	1'30"	3'	6'	12'	24'	48'	1:36'

300	30"	1'	2'	4'	8'	16'	32'	1:4'
225	22-1/2"	45"	1'30"	3'	6'	12'	24'	48'
150	15"	30"	1'	2'	4'	8'	16'	32'

*Days :Hours 'Minutes "Seconds

TAPE LENGTH (FT) ACCOMMODATED ON STANDARD SIZE REELS

TAPE TYPE	OVERALL APPROX. THICKNESS	REEL DIAMETER (INCHES) 3	4	5	7	10-1/2	14
STANDARD	2 mil	150	300	600	1200	2400	4800
THIN	1-1/2 mil	225	450	900	1800	3600	7200
EXTRA THIN	1 mil	300	600	1200	2400	4800	9600

Fig. 6-7. Tape speed vs length chart.

Test for Slitting and Winding

Hold the tape reel against a bright light source, like a window. If the light shines through evenly, the tape is proper. Spotty dark areas indicate an uneven slit. Also, feel the tape pack; it should feel smooth.

Cupping or Curl

These tape defects are illustrated in Fig. 6-1. Cupping is simply tested by feeding the end of the tape over an edge, with the "outside" (base) of the tape resting on the surface. The tape has no cupping if it tilts downward when it is fed past the edge. Curl is detected by unwinding a few feet of tape, holding the reel up and looking down along the free-hanging length of tape. It should be a flat sheet, as illustrated in Fig. 6-1.

Layer-to-layer adhesion is detected by observing how easy the tape unwinds. Hold the reel vertical and slowly turn it; the tape should unwind freely. Any jerky motion caused by sticking to the tape pack will cause wow in many recorders. This defect is particularly noticeable after prolonged storage of a tape. The ultimate test is performed by the **recording and the reproducing** of suitable program material and in comparing the input with the played back signal. There should be no difference in level or frequency response. But this test does require, to be fair, that bias and equalization be properly adjusted. This test is generally referred to as an A-B test.

Magnetic tapes suffer from a weakness called **print-through**, which is found in the thin base tapes, in particular. When the tape is wound onto a reel, adjacent coating layers are seperated by a distance equal to the thickness of the base material. Flux-lines from one layer will reach adjacent layers and under the influence of time, temperature, and external fields, cause a weak "recording" on the adjacent layers. This printed signal can be very annoying in audio recording, while it is seldom observed in video and instrumentation applications. A quiet pause before a loud orchestra opening may, be print-through, contain a faint prelude of the opening. This calls for caution in using thin-based tapes and the use of a conservative recording level.

The **playing time** and the required frequency response dictate the length of tape needed, which should be rounded up to the nearest standard reel configuration. By knowing the recorder's capabilities, the tape speed is established by the required frequency response and the tape length is easily found from Fig. 6-7A, and the reel configuration from Fig. 6-7B.

Instrumentation tapes are high-quality products, designed and manufactured to meet a stringent set of requirements. Quantitative numbers are now attached to slitting tolerance, cupping, layer-to-layer adhesion, abrasion, wear characteristics, sensitivity, response, etc. (While such specifications also exist for the manufacture of audio tapes, these are normally purchased on their established merits, where instrumentation tapes are purchased against a set of specifications which from time to time are checked by the user.) Instrumentation tapes are widely used in the aerospace industry and if a tape for one or another reason fails during an experiment, it can mean the loss of literally hundreds of thousands of dollars worth of scientific data. Thus, the requirements are more stringent.

There are three general classes of instrumentation tape: wideband tapes for recording wavelengths as short as 60 microinches (2 MHz at 120 IPS), medium grade tapes for wavelengths as short as 240 microinches (0.5 MHz at 120 IPS) and standard tapes for 600 microinches (100 kHz at 60 IPS). These tapes are manufactured to meet U.S.A. Federal Specification W-T-0070 (Navy-Ships), which contains a general section and five specific specifications designated W-T-0070-1 through 5. Adherence to these specifications will result in a quality adequate for all requirements from audio through wideband instrumentation recordings. Extreme wideband, low-noise tapes are not yet covered by a specification, since these tapes represent the state-of-the-art products and an area where much research and development is a continuing effort.

These efforts are centered about improving signal-to-noise ratio and a reduction of **dropouts**. A dropout is defined as a 50 percent (or greater) amplitude reduction in the reproduced data. This is a tape error and is generally caused by poor head-to-tape contact during recording and/or playback. A dust particle or a nodule in the coated surface will lift the tape away form the heads and, during recording, this will move the tape away from the otherwise properly adjusted bias signal field and reduce the recorded flux. During playback, it will reduce the signal by the normal spacing loss equal to $54.6 \times d/\lambda$ db, where d is the spacing between the tape and head surfaces. Thus, a dropout during recording will cause a somewhat wavelength-independent signal loss, while during playback it will be most severe at short wavelengths. In instrumentation recording, a good tape should not exhibit more than one dropout per track (50 mils wide) per 100 feet of tape, which roughly corresponds to a dropout every second at a tape speed of 120 IPS.

It is generally possible to recognize a dropout as such in instrumentation recordings and to correct for its influence upon the final data. In computer applications, a dropout is a distinct error and computer tapes are, therefore, thoroughly tested and sold under the warranty of being error free. High-grade computer tapes are certified to be free from any dropouts throughout the reel on any of the 7 (or 9) tracks at standard packing densities.

Computer tapes in actual use are subject to rapid shuttling back and forth through a transport. This imposes the additional requirement that they have a high resistance to wear and generation of dropouts. This is generally tested by a method where a short length of tape is moved back and forth over a head assembly; by monitoring the output signal, it is noted how many passes are required to generate a dropout. This number will generally lay in the vicinity of 100 to 300,000 passes, approaching one million passes for a high-grade computer tape.

It is difficult to test computer tapes, since many factors leave room for proper interpretation of wear data. Elements in the transport such as high-pressure capstan pucks, long sliding support surfaces, and edge guiding constraints are major causes for tape wear and dropout generation. These will naturally differ from transport to transport and the full evaluation of a computer tape for general-purpose use should involve several different tape transports. A recent publication [2] outlines a test combining durability and dropout activity which supplies a comparative ranking of different tapes, without the test being restricted by a requirement for special laboratory equipment. A tape is ranked by just three data points, which are:

1. The number of head feet accumulated when the first dropout is detected as dropout activity.

2. The number of cumulative dropouts which have occurred as the wear length equals 50,000 head feet.

3. The number of head feet accumulated when the first permanent error occurred.

A straight line is drawn joining these three points, as shown in Fig. 6-8. At the point of durability head feet, the line is drawn with a discontinuity of greater slope, signifying an accelerated dropout count due to the catastrophic failure of the tape.

Video tapes must meet the requirements of both instrumentation and computer tapes; they must have a high signal-to-noise ratio and few dropouts. Both affect picture quality, and showers of dropouts are spectacularly annoying.

The testing and evaluation of video tape is ultimately a visual test, and there is a disadvantage that a dropout or a shower of dropouts may occur at the instant where the observer blinks his eye. Methods for testing video tapes have been developed which, on a single graph, will display dropout activity in terms of the number of dropouts and whether they may be visually annoying or not.

An often highlighted feature of helical-scan video recorders is their still-frame capability; that is, with the tape motion stopped, the rotating head assembly will scan a particular portion of the recorded material and display it on the TV screen. A very high temperature is generated in the contact area between the rapidly moving head and the tape (in the order of several hundred degrees C), and this may wear out the tape or cause the formation of debris on the video head (called head clogging). Debris is a general term used for oxide and/or binder buildup on magnetic heads and tape guides. Some tapes have a still-frame capability of only a few seconds

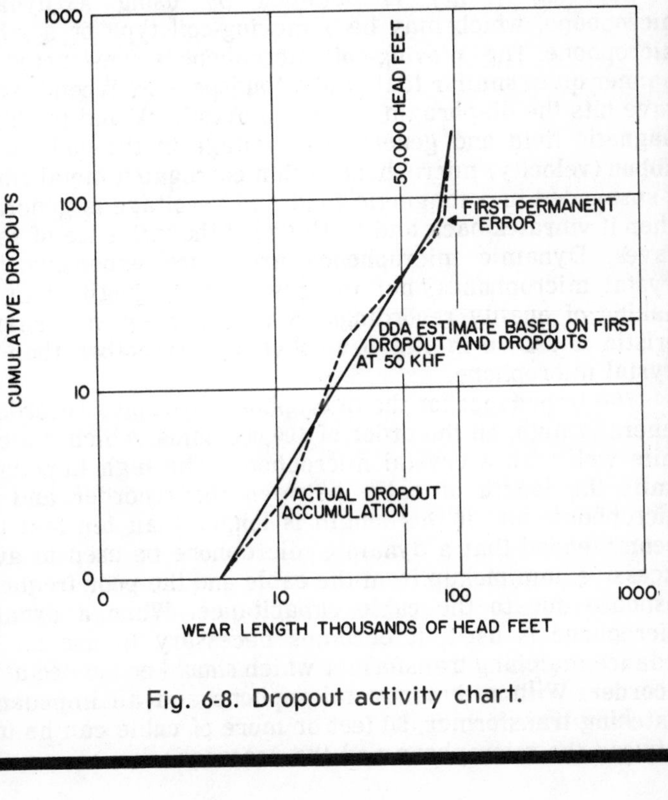

Fig. 6-8. Dropout activity chart.

and a-e obviously rated poor, while others will last in excess of one hour. The still-frame test is consequently used as a figure of merit for a video tape.

MICROPHONES

Tape recorders for home use are generally always supplied with a microphone. Due to price competition, they are just adequate for recordings and typically are of the **crystal** type. In this type of a microphone the diagragm is mechanically connected to the crystal element which vibrates when the sound wave hits the diaphragm. This minute vibration of the crystal generates a varying electrical voltage which is fed via a cable to the microphone input on the recorder. The frequency response of a crystal microphone seldom exceeds 10,000 Hz and is far from smooth, as shown in Fig. 6-9. This poor frequency-response characteristic distorts the original sound and is particularly noticeable when recording music.

Superior quality is obtained by using a **dynamic** microphone, which may be a moving-coil type or a ribbon microphone. The **moving-coil** microphone is constructed in a manner quite similar to that of a loudspeaker. When a sound wave hits the diaphragm the coil moves back and forth in a magnetic field and generates a voltage in the coil. In the **ribbon** (velocity) microphone, a thin corrugated metal ribbon is suspended in a magnetic field and a voltage is generated when it vibrates back and forth under the influence of sound waves. Dynamic microphones are more expensive than crystal microphones, but the cost is fully justified in the making of quality recordings. A typical frequency characteristic is shown in Fig. 6-9, obviously smoother than the crystal microphone.

The **impedance** for the microphone input on a recorder is generally high, on the order of 100,000 ohms, which matches quite well with a crystal microphone. The high impedance limits the length of cable between the recorder and the microphone, and if this length is longer than ten feet it is recommended that a dynamic microphone be used to avoid excessive hum pickup from the cable and the poor frequency response due to the cable capacitance. When a dynamic microphone is used, it becomes necessary to use an impedance-matching transformer which should be located at the recorder. With a dynamic microphone and an impedance-matching transformer, 50 feet or more of cable can be used between the microphone and the recorder.

The three microphone types discussed differ in their directional characteristics as illustrated in Fig. 6-10 where the crystal or moving coil microphones are essentially omnidirectional; that is, they will pick up sound from any direction. This type of microphone is well suited for recording at parties or business conferences and also for recording overall background sounds. The ribbon microphone has a sensitivity pattern that resembles a figure 8; that is, it is most sensitive to sounds coming from its front or back side and it is insensitive to sounds coming in from the sides. By combining the characteristics of moving-coil and ribbon microphones, a cardioid pattern is obtained which essentially makes the microphone sensitive in one direction only. A cardioid microphone is particularly advantageous in avoiding background sound pickup, since it concentrates on a specific sound source. (Chapter 7 contains a discussion on microphone techniques which, in conjunction with the above comments, should assist in selecting the right microphone for the type of recording under consideration.)

ACCESSORIES

Several recorder manufacturers separately offer accessories for their equipment, such as tools or devices that increase the usefulness and application of a recorder.

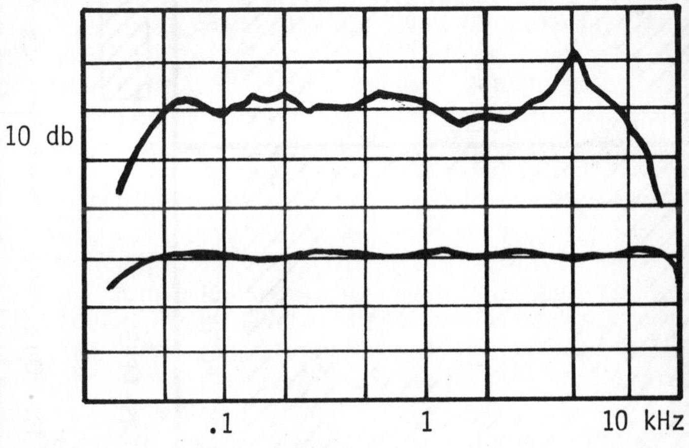

Fig. 6-9. Frequency response curves for a crystal microphone (top) and a dynamic microphone (bottom).

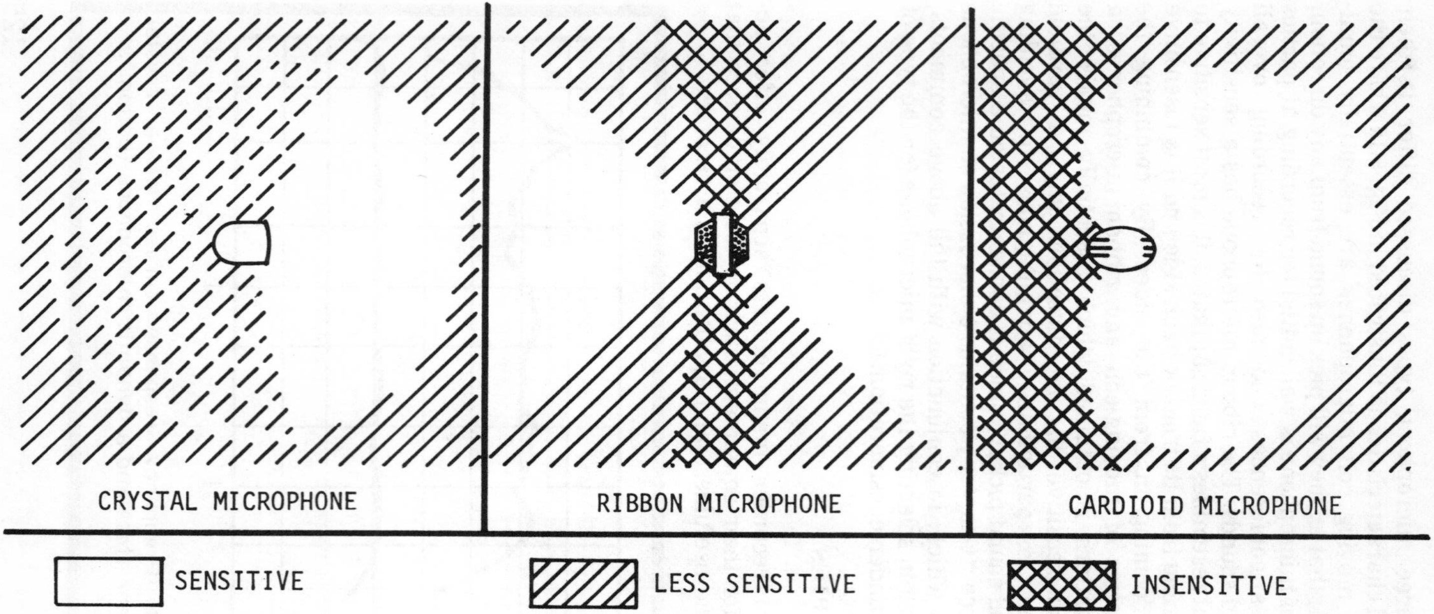

Fig. 6-10. Directional characteristics for the three microphone types discussed in the text.

A magnetic **head degausser**, for example, is an almost mandatory tool to keep the recorder in top condition. The heads (record and playback) in most recorders become slightly magnetized due to switching transients and assymetrical signal waveforms. This causes both distorted recordings and noisy recordings and even a pre-recorded tape can be degraded by additional noise when it passes over a magnetized head. The head degausser is a powerful solenoid with a pencil-shaped core or with two cores forming a narrow gap. It is generally connected to a power line outlet and slowly moved in toward the front of the magnetic heads. Great care should be exercised in not actually touching the heads, since the attraction between the degausser and the core may be quite high, thus causing the degausser to actually hit the head and damage its surface. After holding it in the vicinity of the head for a few seconds, it is slowly moved a couple of feet away from the recorder and the power disconnected.

Other degaussers are used for erasing tapes, which is an important factor in the reusability of tapes. Here again, a variety of degaussers are available. They range from the small unit, which is hand-held and slowly moved in over a reel of tape and moved about it and slowly removed again, to the larger units as shown in Fig. 6-11, which are capable of degaussing the larger size instrumentation tapes.

Another useful tool is a **tape splicer**. A tape may break due to careless handling of the tape transport mechanism or it may be desirable to edit recorded tapes, thereby removing poorly recorded sections such as may be the case with retearsals prior to the recording of an actual performance. The splicer in its simplest form is an aluminum bar with a milled-out groove in which the tape can be laid. The best splice is attained by cutting the tape at a 45 degree angle and then butting the two ends together. Next, a small piece of special splicing tape is used to join the two ends together. Regular glue, cement or adhesive tapes should be avoided under all circumstances. The sticky tape under the pressure of the tape pack will ooze out and interfere with adjoining layers of tape. When such an improper splice goes through a recorder, it may deposit a sticky residue onto the heads which will collect further dirt and thereby cause poor performance. More elaborate splicers with fingers that hold the tape and cut it with a movable arm are also available. A small non-magnetic scissor can be used to cut the tapes in an emergency situation, but it is rather difficult to butt the two tape ends together without any misalignment.

A splice is normally made at a 45-degree angle which allows a smooth transition from one tape end to the other. If

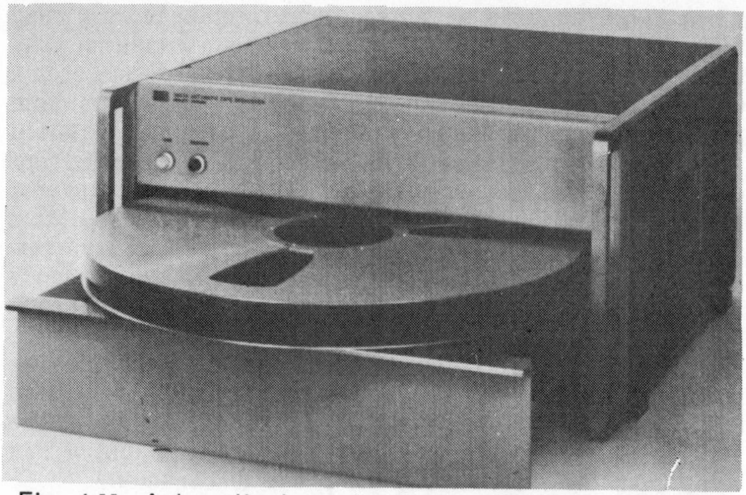

Fig. 6-11. Automatic tape degausser. (Courtesy Hewlett-Packard Corp., U.S.A.)

the purpose of the splicing is to remove a very short annoying sound which may only occupy, say, one-eighth-inch of the tape, the splice should obviously be made at a 90-degree angle. Rather than remove such an error by cutting it out of the tape, it may suffice to put a very short length of splicing tape on the oxide where the annoying sound is located. This will lift the oxide far enough away from the reproduce head to minimize the effect of the click.

In the editing process it is necessary to locate the exact spot on the tape where the splice is to be made. To locate a precise spot a process called **cueing** is helpful and many recorders are equipped with a special control for this function. The tape lifters, if any, are moved to a position that allows the tape to touch the playback head and the two tape reels are moved by hand back and forth. The sound recorded on that particular portion of the tape can then be heard and, with some training and experience, recognized. When the exact position has been found, the tape can be marked right at the reproduce head with a grease pencil. Some splicing devices (such as EDITall, an American splicing and editing block) have a measure that is one and a half inches long. Consequently, if the tape is marked at a position one and a half inches away from the playback head, it will be cut correctly on this splicer, thus avoiding the use of a grease pencil in the immediate playback head area where the slightest deposit of gunk may move the tape away from the head and cause a loss of high frequencies.

As a practical application, editing can provide a very good check on the tape recorder speed accuracy. Connect a tone generator to the recorder and with a full reel, first record the very beginning, then advance the tape almost to the end and again record the same tone. Cut out the beginning and the end of that tape and splice them together, then playback the spliced portion. There should be no tone difference heard when the splice passes over the playback head. If there is a tone change, it is an indication that the tape speed is varying through the reel. This could be caused by too heavy an influence from the reeling arrangement upon the absolute tape speed, which is ideally controlled by the capstan and the capstan puck only. Care should be exercised in splicing various portions of a tape together if this is the case. For critical music listening such a defect will sound like a change in pitch.

Mixers are useful tools in making recordings. These "black boxes" accept several inputs, such as those from microphones, and with individual level controls combine them into a composite signal for the recorder. (The use of a mixer can be appreciated from some of the material in Chapter 7.)

Head phones are very popular for listening to stereo recordings. There are several types on the market and the best way to choose is to try them and listen for the one with adequate sound quality. Listening to stereo tapes via head phones has a strikingly realistic effect and at the same time affords privacy for the listener without disturbing others who may be in the same room. During a recording session, a pair of head phones is an extremely useful device, for the recording engineer can listen to the incoming sound and to the actual playback sound for comparison of the recording quality.

Finally, for use with slide projectors, some manufactures offer **synchronizer** boxes. Speech and music is recorded on one track on a two-channel recorder and the other track contains recorded pulses. Each pulse actuates an automatic slide projector, switching on the next slide. Such synchronizers are also available for movies, too.

References:

1. Ricci, John M.: "Precision Magnetic Tape," Datamation, October 1966, pp. 51-60.

2. Mandle, John B.: "Evaluating Performance of Digital Magnetic Tape," Memorex Monograph No. 5, 1967.

Chapter 7

Applications & Proper Use of Tape Recorders

The operation of any particular recorder is described in its instruction booklet or operating manual which should be studied carefully. A familiarity with the manufacturer's instructions will prevent improper operation and you will be aware of the recorder's capabilities and limitations. However, there are certain fundamental rules that apply to most recorders, and one is on the proper recording level. This applies in particular to music and data recording.

KNOWING THE PROGRAM MATERIAL

It is essential to a good recording that it be free of distortion and that it have a maximum signal-to-noise ratio. All recorders are limited by amplifier and tape noise and the optimum signal-to-noise ratio, therefore, is attained by recording with as strong a signal level on the tape as possible. However, recording at too high a level will cause both harmonic and intermodulation distortion. In music and speech the natural sound is distorted and in instrumentation recording the playback data will be erroneous.

The proper recording level can be attained only with a certain knowledge of the program material and some form of monitoring the recording level, for instance with a VU meter. A magnetic tape's overload sensitivity is essentially the same at all frequencies. The program material, on the other hand, in almost all cases has an uneven amplitude versus frequency distribution and the highest level that can be tolerated is represented by the peak in the amplitude distribution. For example, a male voice is rich in low frequencies, while a female voice is rich in high frequencies. Fig. 7-1 illustrates both the average amplitudes and the peak amplitudes for an orchestra, a pipe organ, and a piano. It is clearly seen that overload during recording of a pipe organ will be caused by

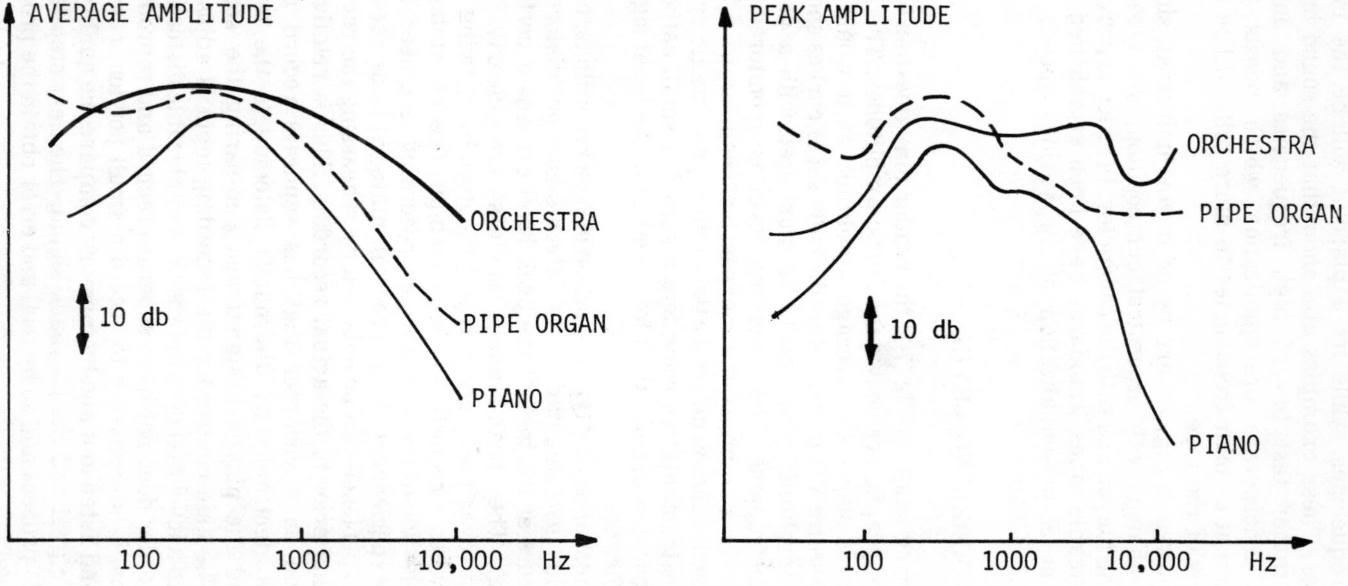

Fig. 7-1. Average amplitudes and peak amplitudes for music programs (after Sivian, Dunn and White, Ref. 1, Chap. 5).

the low frequencies, while for a piano it will be the mid-frequencies. These examples also show that the sound level rolls off rather fast toward high frequencies and audio recorders incorporate pre-equalization which boosts the record current at high frequencies to more fully utilize the capabilities of the tape.

The curves in Fig. 7-1 are by no means universal, since they will change with orchestral composition, the type of music being played and with the conductor. In order to provide interchangeable tapes, standards have been established for the amount of pre-equalization at the various speeds, as discussed in Chapter 5.

RIGHT RECORDING LEVEL

Referring again to Fig. 7-1, the reader may question why two sets of graphs are shown for music amplitudes. The left set of curves shows the average amplitude as it would be measured with a VU meter, while the right set of curves shows the peak amplitude which could be monitored with a peak-reading instrument. The recording level is monitored by various devices, such as the inexpensive neon lamp, a VU meter, a peak-reading meter, or the earlier type "magic eye." In some instrumentation recorders a series of small cathode ray oscilloscopes are installed for monitoring the input signal to each channel.

The most annoying type distortion in tape recordings, both music and instrumentation, is intermodulation distortion which occurs at the instant the peak level exceeds a certain amplitude. The peak-reading devices are clearly advantageous over the average type indicators by showing the operator when a recording level is too high. The VU meter is useful only when the operator or recording engineer has previous experience with the program material to be recorded. This experience can be obtained by rehearsing the record level setting prior to the actual recording. This is relatively easy to do on a recorder that has separate record and reproduce electronics by alternately listening to the input signal and the playback signal and comparing the sound quality. The gain control for the recording input is adjusted upwards to a level that does not result in intolerable distortion. If the recorder does not have separate record and reproduce electronics, it is necessary to record a small portion, rewind the tape and listen to it and by memory compare the quality of the input signal and the playback signal. The next step is to repeat this process and so by trial and error obtain the proper record level.

Orchestral music often has a dynamic range that exceeds that of the recorder's and, in this case it is common practice to "ride gain." The recording engineer will be familiar with the program and can anticipate loud passages and very slowly reduce the gain prior to such a passage. This requires great skill and experience and should not be attempted by a novice. He would be better off with a recorder with an automatic gain control. Such a unit can be purchased as an accessory for the recorder. But the true value of an automatic gain control is attained only if, upon playback, the AGC action is reversed, which in turn requires an additional control track on the tape. Recorded on the control track is a signal that represents the AGC action and upon playback is fed to the control of another AGC unit. This method has found little commercial application.

The rapid rise of distortion versus input level is illustrated in Fig. 7-2. There is only a 12 to 14 db margin between the 1 percent harmonic distortion level and full saturation of the tape. Intermodulation distortion, as a rule of thumb, is three times higher than the harmonic distortion level.

MICROPHONE TECHNIQUE

The quality of a recording depends heavily on room acoustics, since the microphone will pick up the **direct sound** from the sound source as well as the **indirect sound** coming from reflections off the walls in the room (reverberation). The contribution from the indirect sound plays a major role in the recorded sound quality since it "colors" the sound. A recording made in a well-damped room (carpets, upholstered furniture, drapes, etc.), where the reverberation is small, will sound dry and unnatural. A recording made in an empty room with hard walls, on the other hand, will contain a large amount of reverberation, which in speech will mask the intelligence and in music will contribute heavy echo effects.

The ratio between the direct sound and the indirect sound can be varied by changing the distance between the sound source (for instance, a speaker) and the microphone. It will be necessary to experiment to find the best microphone position, where the ratio between the direct and indirect sound is the best possible and the recording sounds most natural. The reader is encouraged to undertake a simple experiment to observe this phenomena:

Make four consecutive recordings, each time repeating the same sentence. Position the microphone eight inches from your mouth during the first recording, one yard during the second recording, two yards during the third, and four yards

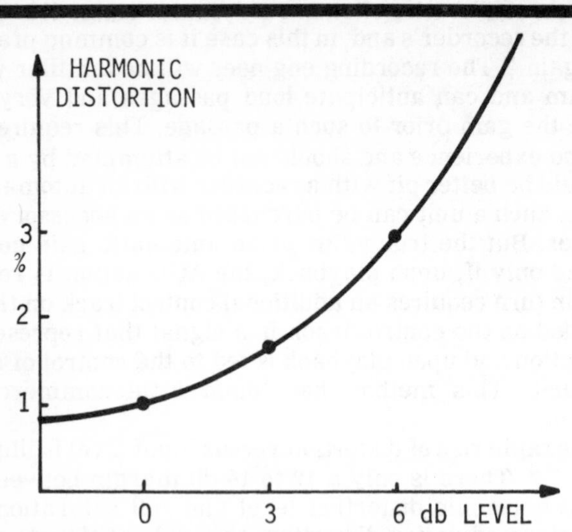

Fig. 7-2. Harmonic distortion vs record level. (Ref. normal recording level at 1 percent distortion.)

during the fourth recording. In each case the record gain control on the recorder must be adjusted for a normal record level (increased each time). Now rewind the tape and play back the four recordings. You will notice a striking difference in the sound. The first recording will sound dry and unnatural; the next recording will be much more alive, and at the same time the listener gets a sense of the presence of the room the recording was made in. During the third and the fourth recordings, you will not only get the impression that the distance to the microphone has been increased but that the room also seems to have grown larger. This simple experiment clearly shows the importance of a proper microphone position to preserve the natural sound and to emphasize or de-emphasize the acoustical quality of the recording room.

An additional effect is obtained by using the microphone's directional characteristics. If an omni-directional microphone is used in an acoustically poor room (large degree of reverberation), it becomes necessary to position the microphone near the speaker. A cardioid type microphone is preferable in this case. If the recording is planned for a small orchestra group, the microphone position becomes more difficult, since different instruments produce different levels. It may in this case be necessary to try not only several microphone positions but also to place the musicians in various positions. A better solution may be to employ two or more microphones and use a mixer to obtain proper balance between the microphone inputs.

The following rules applied with reasonable flexibility will result in good recordings:

Speech: Place the microphone at a distance of 12 to 24 inches from the speaker. If any "s" sounds are annoyingly loud, turn the microphone slightly, as shown in Fig. 7-3. When recording a chorus where individual voices may have varying levels, the singers should be spaced from the microphone as shown in Fig. 7-3. If an intimate effect is required, the speaker should be close to the microphone and talk in a soft voice. The impression of a large room is obtained by placing the microphone away from the speaker.

Piano music is best recorded with the microphone positions shown in Fig. 7-4. An alternate technique is to place the microphone very close to the sound board in a grand piano. Piano music accompanied by song is more difficult to record. The piano will normally have a louder sound than the singer and the choice is either to use two or more microphones with mixers or to position the singer between the microphone and the piano, as shown in Fig. 7-4. If the piano player is singing along to the music, the microphone should be placed next to the player, never on the piano, since the microphone may pick up vibrations from the piano.

Pipe organ recordings place a great demand on both the microphones and the tape recorder. It is advisable to place the microphone at a rather large distance from the organ so as to obtain a good balance between the direct and the indirect sound. Caution should be exercised in selecting a moderate record level, particularly with recorders having NAB pre-equalization which boosts the low frequencies and easily causes intermodulation distortion.

Orchestral music is in general recorded by two general techniques. Symphonic music may be recorded with one microphone positioned over and behind the conductor. Dance or popular music is recorded with several microphones which are mixed in a suitable ratio to give the desired effect.

Stereo recordings are made with two microphones; the output of each is recorded on separate tracks. By playing back these two tracks through separate speakers, a special effect is obtained which gives the listener a sound picture, as if he were placed in front of the orchestra. This feeling is obtained by the phase difference between the sound from the two speakers and

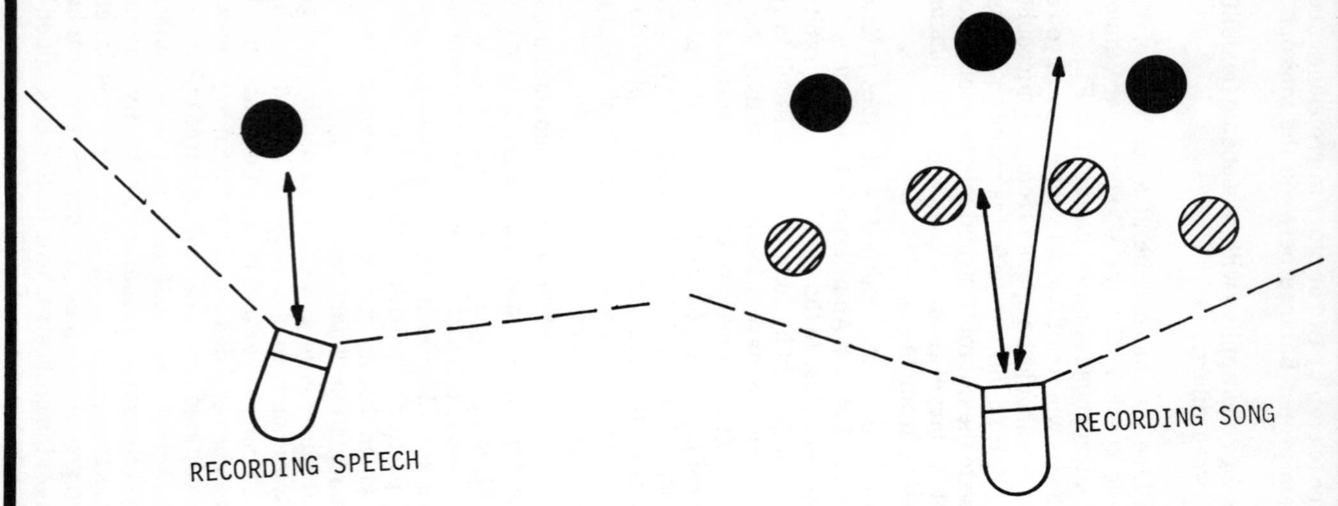

Fig. 7-3. Microphone placement to record speech and song.

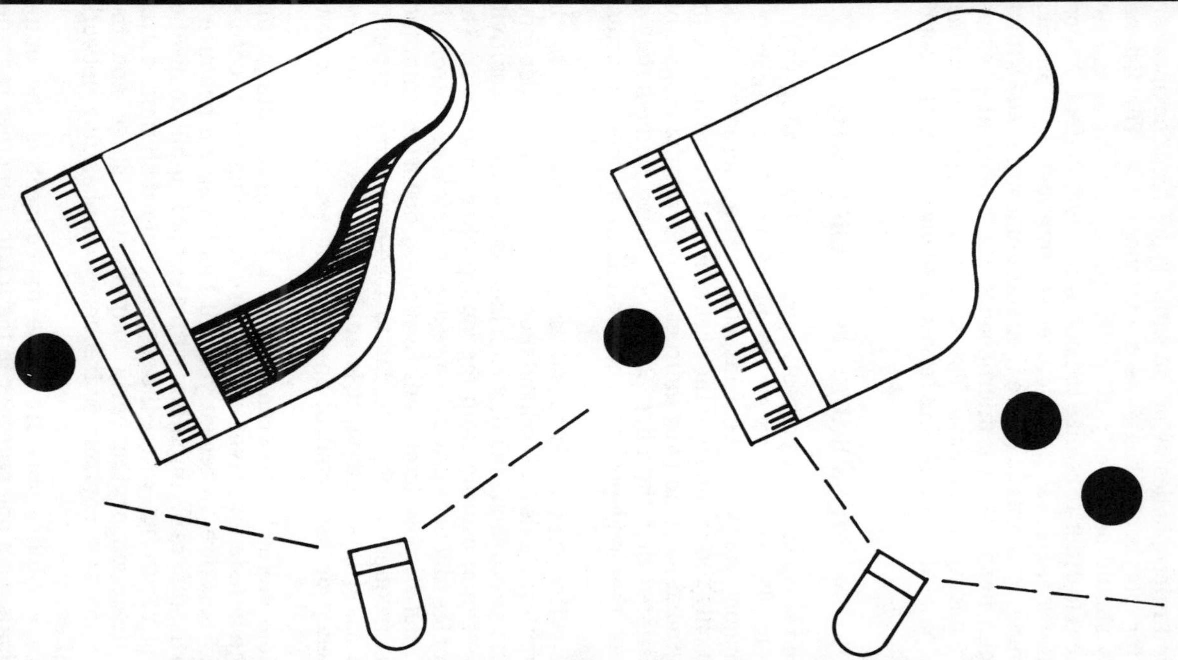

Fig. 7-4. Microphone placement to record a piano and piano and vocal.

also by the intensity of the sound. The amplification, gain settings and frequency response of each channel must be identical (plus or minus 1 db), and this demands that the selected microphones be of the same type. Also the speakers used for playback should be a matched pair. The listener ideally should be exactly at the midpoint in front of the two speakers in order to realize the optimum stereo effect. A more realistic spacial sound picture is obtained by **binaural** recording, a technique where the recordings are made with an artificial head—the two microphones are placed at positions corresponding to the ears. The playback of such recordings via a pair of head phones gives strikingly realistic sound results.

RECORDING FROM RADIO, TELEVISION, ETC.

The simplest (but poorest) method for recording from a radio or TV set is merely to place the tape recorder's microphone in front of the loudspeaker. The microphone picks up not only the sound from the program material, but also the background noise and any indirect sound from the room. This necessitates that the microphone be placed immediately in front of the loudspeaker, but even then a poor recording results.

A better method is to solder or clip two wires to the speaker terminals and connect them to the tape recorder input (not the microphone input, which has far too much sensitivity for the signal coming from the loudspeaker terminals). This eliminates any background noise and indirect sound from the room, but it most likely will result in a distorted amplitude versus frequency response due to tone-correcting networks. When using this method, it is advisable to see that the low-frequency or base control on the radio set is turned down somewhat.

Both methods are accompanied by a disadvantage; any change in listening level will require resetting the recording level. Therefore, in recent years it has become a practice to equip the tuners or amplifiers with a special output connection for tape recording, where the signal is tapped off immediately after detection or the early amplification stage and thus before any gain controls or frequency emphasizing networks can affect it.

Tape duplications are made by connecting the output terminals from one recorder to the input of another recorder. The best sound quality is obtained by playing the tape to be re-recorded backwards. Simply play the tape through, then switch takeup and feed reels. Now, the tape will be playing backward as it is duplicated on the other machine. This technique will not work with dual- or multi-track recorders, though. There are minute phase shifts inherent in all tape recordings and these can be reduced by the backward duplication process. The backward process should be of little concern to the user, since it is a good rule to leave the tape on the take-up reel after playing it and then to store it in that fashion. This prolongs the tape's life, minimizes dust pickup and edge damage.

SOUND EFFECTS

The advent of two-channel recorders for stereo recording and playback allowed a couple of extra features to be incorporated into the art at practically no extra cost. One of these is the **echo** effect which is achieved by connecting the playback head output to the record head input thereby re-recording a recorded signal to obtain an echo effect. (This should not be confused with reverberation, since the echoes are distinct and often spaced several hundred milliseconds apart.)

A more interesting and useful feature in many of these machines is the **sound-on-sound** technique. This is useful, for instance, in the recording of piano music accompanied by a vocal. A recording of the piano is first made on track one and possibly repeated until the quality and performance is satisfactory. The sound-on-sound selector switch is now turned on and the vocal is recorded. When the tape is played back, the first recorded piano music can be heard using a set of head phones, but now accompanied by the vocal. The singer is recorded on track number two and it can be repeated until the vocalist is satisfied with the quality. It is possible to add, for example, a guitar by flicking the sound-on-sound switch back to its original position and the recording will be made on track number one of the mixed input from the guitar microphone and the piano plus the song recording on track number two. This feature has many possibilities and may be used in multiple guitar recordings (as employed by Les Paul in the 1950s).

Sound effects can be dubbed in, too. They can either be purchased ready recorded on special records, or can be "homemade," as listed below. In most cases, the microphone should be placed close to the sound image:

Breaking through a door: Break a thin piece of wood with a hammer stroke.

Car brakes: Move a metallic item (for example, a fork) over a glass sheet.

Car crash: Let a few metal sheets fall onto the floor.

Car door: Close a thick book.

Elevator: Start (or stop) a vacuum cleaner.

Fire: Slowly squeeze cellophane wrapping in front of the microhpone.

Horse steps: Cut a coconut in half and clap the two halves against each other. (Cover one of the halves with, say, a sock if the sound should be of horses riding on a wooden bridge.)

Jet plane: Hold a hairdryer in front of the microphone and let the airstream howl over a sheet of cardboard.

Locomotive: Glue two sheets of abrasive paper onto two pieces of wood; rub them against each other.

Machine gun: Squeeze a piece of aluminum wrapping foil.

Ocean liner: Blow downwards into an empty bottle.

Power boat: Hold the microphone close to a household mixer.

Rain: Shake 20 to 30 dry peas in a net sieve over a microphone, or let a handful of sand slowly drizzle onto a sheet of paper or cardboard.

Rifle: Slap a flat ruler against a table surface.

Rowboat: Move a couple of pieces of wood in and out of a bucket filled with water.

Steps: Squeeze a sheet of paper in a rhythmic fashion.

Steps on snow: Squeeze a handful of starch in a rhythmic fashion.

Surf: Brush a metal (or cardboard) sheet in a rhythmic fashion with two brushes.

Thunder: Hold a metal sheet by one corner and shake it violently.

Voice over a telephone: Speak into a paper cup.

Waves: Fill a tray with water and move a hand around in the water. Hold the microphone close by.

Wind: Move a piece of silk over a table edge or imitate the sound by whistling through your teeth.

The above suggestions should provoke your imagination to devise many more.

Chapter 8

Care & Maintenance

A tape recorder, like other instruments (for example, a motion picture camera), is subject to malfunctions that are best prevented by proper maintenance. And a reel of magnetic tape (like a roll of photographic film) is sensitive to handling, storage, and shipment hazards. Routine maintenance of the recorder and the observance of a few fundamental rules about tape handling will assure the user of better performance and longer life for both equipment and tapes.

RECORDER MAINTENANCE

Cleanliness is fundamental for the proper operation of any magnetic recorder. Dust particles will not only cause dropouts on magnetic tapes, but may also shorten the life of the recording equipment, especially the heads and bearings. Good cleaning rules apply to home sound recording equipment, but the demands are even greater for instrumentation, video, and computer recorders. Ideally, these should be operated in a clean-room atmosphere to avoid the accelerated generation of dropouts. Many professional facilities also have restrictions on the use of food and tobacco, since tobacco ashes can easily accumulate on the tape.

Any part touching the magnetic tape on its pass through the transport should be cleaned at regular intevals–tape guides, heads, capstans and rubber pucks. If this practice is not followed, dirt can accumulate on the tape guides and heads and act as an abrasive agent and scrape the oxide coating off the tape. Soon the oxide buildup from the scratching may break away and be redeposited elsewhere on the tape. As the tape is wound tightly onto the take-up reel, any loose oxide may be firmly embedded into the tape surface and cause dropouts the next time the tape is used. Care should be exercised in the selection of a **cleaning solution**, since some agents may do more harm than good. About the only cleaning solution recommended for video-tape equipment is Freon TF. The reason is that Freon TF will flush off oxide particles and debris without softening the oxide or the backing. Also, Freon TF will not damage the rubber capstan idler. Fig. 8-1 states the effects various cleaning solutions generally exhibit.

Another trouble source—an easy-to-control but often neglected problem—is head magnetization. A magnetized record head increases second-order harmonic distortion and the overall noise level. Therefore, a small head degausser should be used at regular intervals (for example, every eight running hours).

With regard to lubrication and head alignment, refer to your recorder manual. Many recorders have bearings that are lubricated for life. But, if lubrication is required in certain parts of the recorder, care should be exercised not to spill any oil on the capstans, rubber pucks, or other sensitive areas.

HANDLING TAPES

Many magnetic tape problems can be avoided if the user follows a simple bit of advice: **Do not rewind the tape after recording or replay**, but store it immediately in its container, standing on end. Numerous tape failures and dropouts are the result of not following this practice. When winding or rewinding a tape, the recorder may produce an uneven "pack" or "wind" with protruding layers which are subject to damage in handling. By holding the reel the flanges are quite often squeezed against the tape and protruding layers may be nicked, torn or permanently deformed.

A transport winding mechanism may also wind the tape without sufficient tension. Later handling will then cause the pack to shift from side to side against the flanges, leading to later edge damage. A loose pack is also subject to tangential slippage between layers, called cinching. Cinching is likely to occur in a reel of tape with one or more regions of too low tension, especially if subjected to a rapid angular acceleration of deceleration which occurs during the starting and stopping of a tape handler. Cinching is shown in Fig. 8-2. During such slippage the tape may actually fold over on itself so that permanent creases form immediately or perhaps later when tension is applied and the tape attempts to return to its original position. Creases cause dropouts by introducing a separation between tape and heads.

If the tape is left on the recorder or placed on a shelf outside of its container, as is often the case, dust will collect on the tape within a very short time. When the tape is later played back, dust particles will cause dropouts, permanently

CLEANING SOLVENT	HEALTH HAZARD	EFFECT ON VIDEO MAGNETIC TAPE	FLAMMABLE	EFFECT ON RUBBER
FREON TF	VERY SLIGHT	NONE OR NEGLIGIBLE	NO	NONE OR NEGLIGIBLE
ACETONE	VERY SLIGHT	SOLUBLE	YES	NONE OR SLIGHT
CARBON TETRACHLORIDE	GREAT	NEGLIGIBLE	NO	SLIGHT
ETHYL ALCOHOL	VERY SLIGHT	NEGLIGIBLE	YES	NONE OR SLIGHT
HEPTANE	SLIGHT	SOFTENS	YES	SWELLS
METHYL ALCOHOL	SOME	NEGLIGIBLE	YES	NONE OR SLIGHT
NAPHTHA	SLIGHT	SOFTENS	YES	SWELLS
MEK	SOME	SOLUBLE	YES	NONE OR SLIGHT
TRICHLOROETHYLENE	SOME	SOLUBLE	NO	SLIGHT
XYLENE	SOME	SOLUBLE	YES	SWELLS

Fig. 8-1. Table of tape cleaning agents.

Fig. 8-2. Cinched tape.

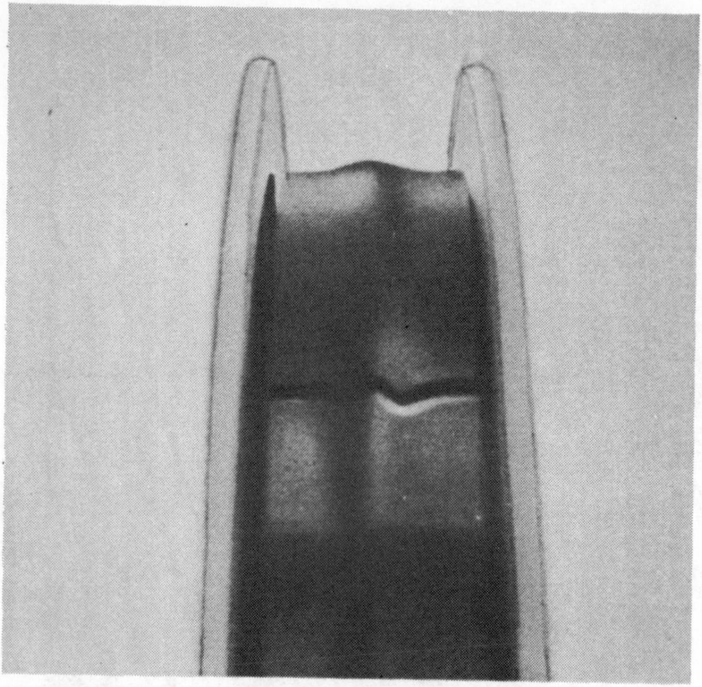

Fig. 8-3. Ridged tape.

damage the tape, and may even scratch the magnetic heads. Dust particles may again combine with debris from the tape and deposit it on the guides and heads. These protruding particles will scratch the tape surface and further aggravate the dropout situation. There also may be a continuous scratch in the tape surface or backing throughout an entire reel. The slight ridge thus produced can multiply through the layers of tape wound on the reel to produce a much larger ridge in the outer layers of the tape--often large enough to cause permanent deformation in the outer layers. For instance, a ridge only ten-millionth of an inch in height will, if continuous throughout the tape that is wound on a standard 14" reel, appear as a 30 mil high ridge at the outside layer. This is shown in Fig. 8-3.

A final word of caution applies in particular to video tapes. Hands and fingers have body oils and salts which attract foreign particles. Each time a tape is threaded on a video transport, the operator may transfer these oils and salts to the tape or to the tape guides. This, naturally, is a problem whenever a video tape is spliced. During a normal splicing or editing session, the tape must necessarily be handled a great deal. At many video tape facilities, white cotton gloves are worn, which helps to reduce this problem considerably.

In order to make a perfect video splice during an editing session, a solution such as "EDIVUE" is used. When this device is placed, in its holder, over the tape a grey powder remains on the tape and the editor can "see" the edit pulses that have been recorded onto the video tape. Unless this powder is removed from the tape completely, dropouts are inevitable.

For storage of tape, the following rules apply:

1. Tape should always be stored in its container with the reel on edge rather than in a flat position. This will tend to eliminate the sideways shifting of the pack against the flanges.

2. Tape should be stored under controlled environmental conditions. It is desirable to maintain the temperature between 40 and 90 degrees F and the relative humidity between 20 and 80 percent. Further, large or sudden changes in environment should be avoided.

3. Tape which has been stored under less than ideal environmental conditions should be conditioned by allowing it to remain in a suitable environment for at least 24 hours prior to use.

4. When large changes in temperature cannot be avoided, the probability of damage to the tape can be minimized if the reel hub has a thermal coefficient of expansion similar to that of the base film. Most plastic reels have a thermal coefficient about twice that of the polyester base film, while the thermal coefficient of aluminum is very nearly equal to that of polyester.

During shipment of a tape, mechanical agitation will tend to shift the tape pack, especially if wound under improper tension. Any abrupt temperature change during shipment should be avoided and this is best done by placing the reel in special fiber-board shipping containers. This will also protect the reel flanges from being bent to a point where the edges of the tape rub on the flange.

Stray magnetic fields may cause some degree of erasure of the information recorded on the tape. There are a few cases on record where tape has been completely erased during shipment and if such fields are known to exist, special shielding containers are avilable.

TROUBLESHOOTING THE RECORDER

The following troubleshooting guide is applicable to most recorders. However, it is necessary that the operating manual and/or the service manual be consulted for a particular recorder prior to repair. For troubleshooting the recorder electronics, instruments similar to those used for amplifiers are normally required (tone generator, VTVM, VOM, oscilloscope). It is also useful to have a flutter meter and test or alignment tapes which have signals recorded for alignment of the reproduce head, setting of the record level indicator, frequency-response tests, and a tone for wow and flutter tests. To clean various parts in the recorder, it is useful to have on hand a proper cleaning fluid and Q-tips (see Fig. 8-1). Listed below in tabular form are typical troubles, their possible causes, and steps for their correction:

ELECTRONICS PROBLEMS

TROUBLE	POSSIBLE CAUSE	CORRECTION
Noise	DC magnetized heads	Degauss the magnetic heads with a suitable head degausser as outlined earlier.
	Input stage	A faulty resistor will cause noise in the input stage and should be replaced. Faulty electrolytic capacitors may also cause noise and should be replaced. Also check for noisy vacuum tubes or transistors.
	Distorted AC bias	Check the oscillator and bias amplifiers for proper, quiet operating voltages and waveforms.
Hum	Faulty shield or ground connections	Ground and shield connections may become corroded and should be scraped clean.
	Open-circuited reproduce head	Replace the reproduce head.
	Poor power supply decoupling	Replace faulty electrolytic capacitors.

Distortion	Tape overload	Carefully monitor the record level. If the overload persists, check the setting of the level indicator on the recorder, which is easiest done with a test tape. For details, see Chapter 10.
	No or too little AC bias	Check the oscillator amplifier. Also check that the proper tape, for which the bias level has been set, is used on the recorder.
No or poor erasure	Faulty bias oscillator	Check the oscillator for proper operation. If the oscillator has been malfunctioning, this would also have resulted in as high a distortion level as above.
	Debris on erase head	Debris will lift the magnetic tape away from the erase gap. The erase head should be cleaned with a proper solvent and a Q-tip.
Poor frequency response	Debris on heads	Clean all magnetic heads with a suitable solvent and Q-tips.
	Misaligned heads	First check the alignment of the reproduce head using a test tape. Next, check the alignment of the record head by recording a high-frequency note and then adjust the record-head azimuth screw for maximum output level. The azimuth adjustment should be made only by a person familiar with the recorder.
	Faulty equalizer	Poor contacts in the equalizer switching circuit may cause the equalizer to function improperly. Faulty components may likewise cause this problem.

	Skew	This condition will normally manifest itself as a variation in the high-frequency output level and is most likely to be caused by a worn or misaligned capstan rubber idler. A new idler should be installed and/or realigned.
	Smeared head gaps	Foreign particles (for example, dust) may cause excessive scratches on the head surfaces which will cause the material to cold flow across the gap and thereby destroy otherwise parallel gap edges. This phenomenon also may appear if the mu-metal cores wear away faster than the head shell; a thicker tape will then no longer conform to the head contour and make proper contact with the gap.
	Wrong tape	Tape from different manufacturers requires a slightly different bias setting for optimum performance. Once selected, the same type tape should be used in future recordings.
No recording	Faulty amplifier	Follow normal amplifier troubleshooting procedures. The record-head current is normally referred to in the service manual and can usually be measured across a 10-ohm resistor in the ground leg of the record head.
	Faulty record head	An open record head will result in no voltage across the 10-ohm resistor. (A short-circuited record head or cable connection will allow current to flow through the 10-ohm resistor and should be checked very carefully.)

No playback	Faulty amplifier	Troubleshoot the reproduce amplifier.
	Faulty reproduce head	This is easy to verify; run a pre-recorded test tape through the recorder and check for proper output levels. A shorted reproduce head will result in no output voltage, while an open reproduce head will introduce excessive hum.

TRANSPORT PROBLEMS

TROUBLE	POSSIBLE CAUSE	CORRECTION
Wow	Debris on capstan	Clean the capstan and rubber puck with a suitable solvent.
	Damaged rubber puck	If power is shut off to the recorder while in the play mode, the rubber puck will remain engaged against the capstan and indent the rubber puck. Such an indentation may be removed by letting the recorder run in the play mode several hours without tape. Otherwise, the rubber puck must be replaced.
	Worn belts, pucks, bearings	Replace worn-out parts.
	Tape scraping on reel flanges	Rewind the tape onto a new reel that has no bent or damaged flanges. Also make sure that the inside edges of the reel flanges are free from any nicks or scratches.
	Heavy oil and dirt in bearings	Clean all bearings as outlined in the service manual and relubricate.
Too slow speed (drift)	Reel tension too high	Adjust reel tension in accordance with the service manual instructions.
	Debris on rubber puck	Debris may cause a rubber puck to become excessively smooth, in which case it should be cleaned with a suitable solvent.

	High bearing friction	Clean and lubricate all bearings as described in the service manual.
	Speed control error	Many portable recorders and most instrumentation recorders maintain their speed accuracy by an electronic servo system. There may be several causes for malfunction in a servo system and the reader is referred to the service manual for the particular recorder.
Squeal (distortion)	Debris on heads and guides	Any disturbance in the tape path through the recorder may cause excessive scrape flutter. Heads and guides should be cleaned with a suitable solvent.
	Excessive tape tension	Adjust the tape tension devices in accordance with the service manual.
	Worn felt pads	Replace.
	Excessive felt pad pressure	Excessive felt pad pressure, either against a guide or against a head, will cause excessive scrape flutter. Adjust the felt pad pressure in accordance with the service manual or by successive experiments.
No tape motion	Broken belts or damaged mechanical parts	Replace damaged parts.
	Blown fuse	Check all fuses and replace if burned out. If a fuse repeatedly burns out, the recorder should be overhauled by a service shop.
Tape breakage	Worn brake pads	Replace worn brake parts and adjust in accordance with service manual.
	Maladjusted brakes	Readjust brakes in accordance with service manual.
Tape throws a loop	Maladjusted brakes	Adjust the brake system in accordance with the manual instructions.

References

1. Eldridge, D. F.: "Causes of Failure in Magnetic Tape," Memorex Monograph, No. 3, May 1964.

2. Reynolds, K. Y.: "Causes of Failure in Broadcast Video Tape," Memorex Timelytopics, No. 1, December 1966.

Chapter 9

Specialized Techniques

The preceding chapters have been devoted to basic analog magnetic recordings, with only casual references to digital recordings. Now, let's discuss several variations in the schemes of digital recording and playback. The resolution of the analog magnetic recording technique is presently limited from a few Hz to approximately 20,000 Hz per linear inch. Playback of very low frequencies requires very low tape speeds and rapidly becomes impractical due to the resulting poor signal-to-noise ratio and the high rate of flutter. Flutter is the ultimate limit when a stick-slip motion of the tape occurs (at speeds normally less than .001 IPS). Fig. 9-1 shows how flutter varies with speed in an otherwise well-designed tape transport. Stick-slip is a friction phenomenon where the tape stops, is stretched by the capstan force and then moves in a jerk to the next halt (similar to moving a finger down a humid glass surface).

High frequencies are recorded and reproduced at high tape speeds. The limiting factors are now high-frequency losses and the formation of an air film between the tape and the magnetic heads. The tape tension must be increased to avoid spacing loss. Present state-of-the-art recorders operate at 2 MHz with a tape speed of 120 IPS. Recording and playback down to a DC signal is necessary in several applications and the upper 2 MHz is unsatisfactory for quality television recording. Fig. 9-2 indicates the necessary frequency ranges for several uses. While the recording and playback of speech and music has been covered, we shall now direct our attention to other forms of data recording, what the nature of these data are, and what methods are used to record and play them back.

OCEANOGRAPHY AND GEOPHYSICS

One branch of oceanographic research concerns itself with the measurement and recording of ocean water tem-

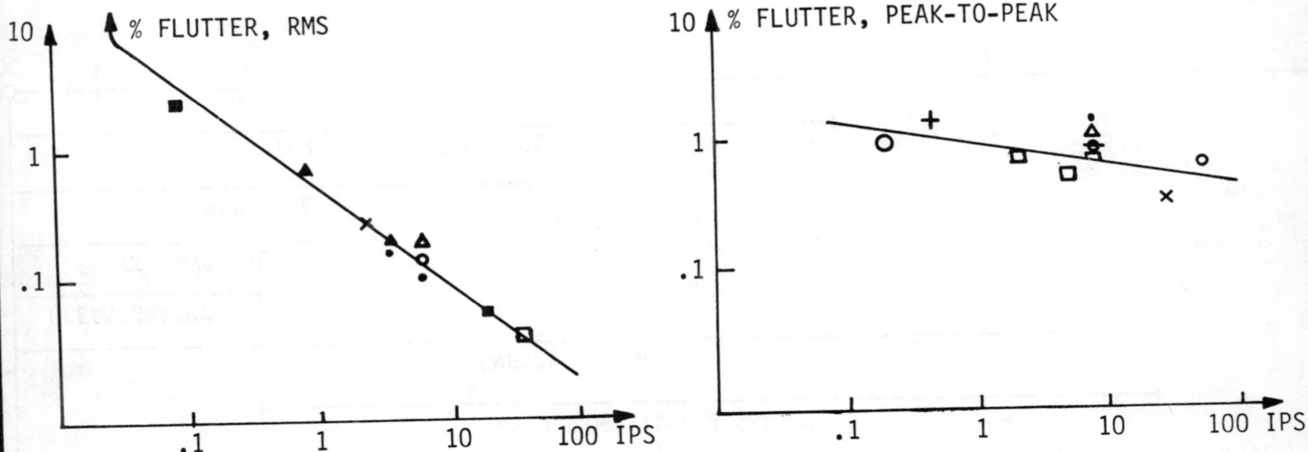

Fig. 9-1. Flutter vs tape speed curves. (Courtesy W. B. Snow, Santa Monica, Calif.)

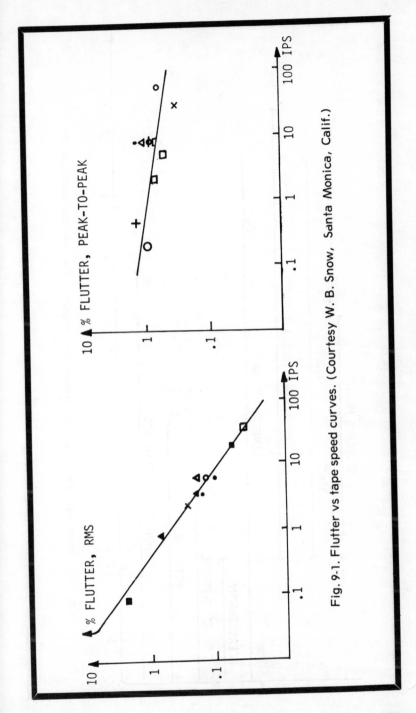

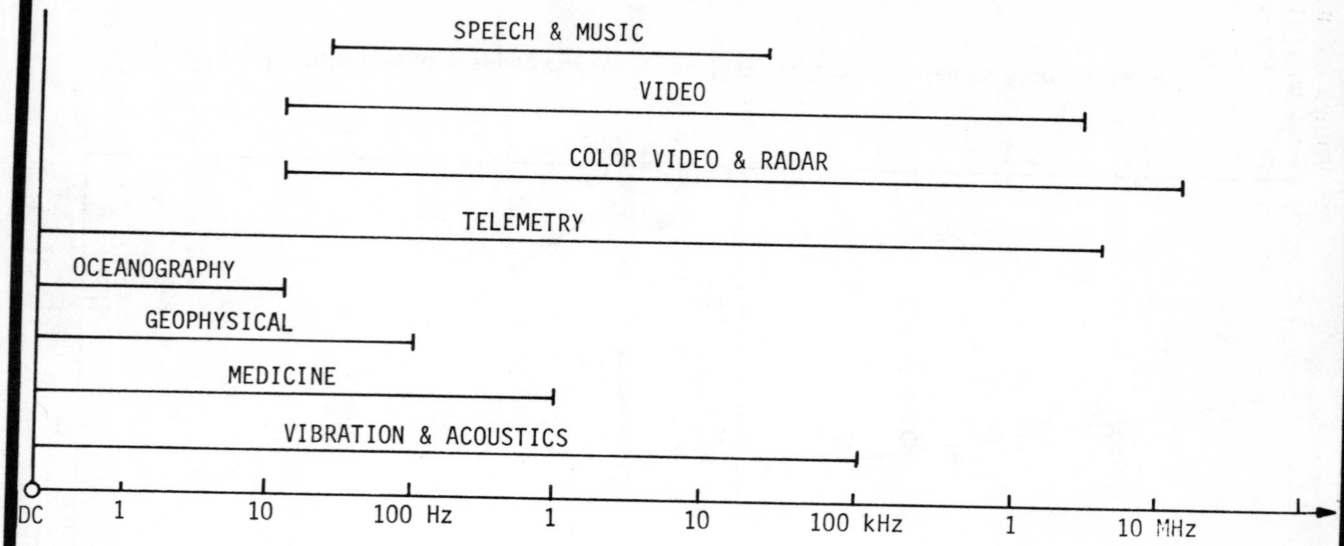

Fig. 9-2. Required tape recorder frequency response for various applications.

peratures, salinities, changes in currents, etc. These are all slowly varying phenomena and, therefore, require only a few cycles of bandwidth, extending down to direct current. In geophysics, measurements are made of earth movements, which may be slow creeps along an earth fault, measured with pressure transducers, or sharp earth tremors, which are measured with seismographs. Earth movements occur so slowly that recordings must continue for weeks; therefore, the high packing density of magnetic tape is an attractive tool.

Magnetic recording down to direct current is no problem in itself, since it merely requires a direct-coupled record amplifier. The limitations are in playback where the physical size of the reproduce core and the slowly varying flux changes reduce the playback voltage to a value below system noise.

It is possible to increase the low-frequency output by the use of a Hall-effect element. This is a semiconductor material that, under the proper bias conditions, will produce an output voltage proportional to a magnetic field. For instance, it can, be inserted in the back gap of the reproduce head, as shown in Fig 9-3. Although a Hall element will produce an output voltage when in a DC field, its application in the magnetic core structure is still limited by the long wavelength limitation of this structure. Second-order harmonic modulators have also been used to extend the low-frequency range of the reproduce

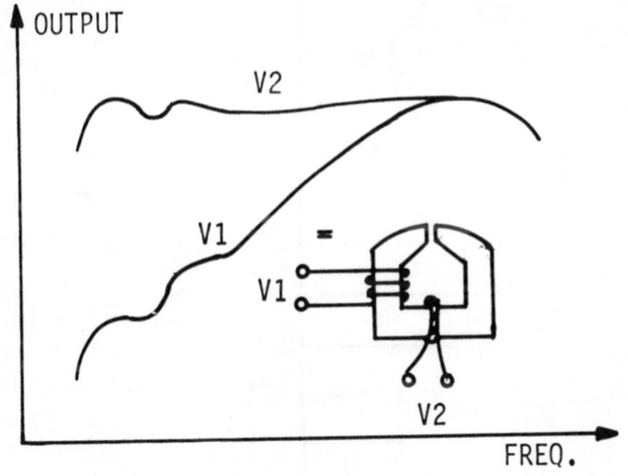

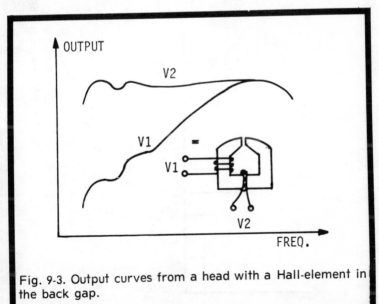

Fig. 9-3. Output curves from a head with a Hall-element in the back gap.

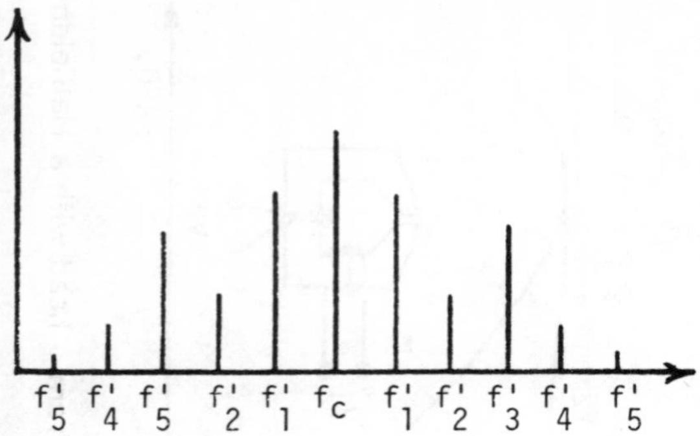

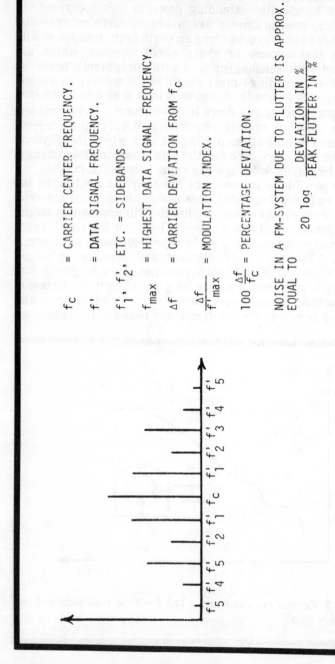

f_c = CARRIER CENTER FREQUENCY.

f' = DATA SIGNAL FREQUENCY.

f'_1, f'_2, ETC. = SIDEBANDS

f_{max} = HIGHEST DATA SIGNAL FREQUENCY.

Δf = CARRIER DEVIATION FROM f_c

$\frac{\Delta f}{f'_{max}}$ = MODULATION INDEX.

$100 \frac{\Delta f}{f_c}$ = PERCENTAGE DEVIATION.

NOISE IN A FM-SYSTEM DUE TO FLUTTER IS APPROX. EQUAL TO

$$20 \log \frac{\text{DEVIATION IN \%}}{\text{PEAK FLUTTER IN \%}}$$

Fig. 9-4. FM frequency spectrum and basic definitions.

head[1] and a tape reproducer system has recently been demonstrated where the response is truly extended down to Direct Current[2].

Low-frequency recording and reproduction was introduced earlier by using modulation techniques. These are either frequency modulation, pulse modulation, or analog-to-digital conversion techniques. In **frequency modulation** recording a voltage-controlled oscillator is added to the record circuitry. Any change in input voltage will change the frequency of the oscillator and this change will represent the input voltage. The output from the oscillator can be expressed mathematically:

$$v = V_o \, [J_o(\beta) \cos \omega_o t$$
$$+ J_1(\beta) \cos(\omega_o + \omega_1) t - J_1(\beta) \cos(\omega_o - \omega_1) t$$
$$+ J_2(\beta) \cos(\omega_o + 2\omega_1) t - J_2(\beta) \cos(\omega_o - 2\omega_1) t$$
$$+ \text{-------------------------}]$$

where J (beta) is a Bessel function, and beta equals delta f/f, the modulation index; f is the quiescent oscillator frequency and f1 is the data frequency.

This formula shows that the modulated signal (carrier) has an upper and lower set of sidebands, and the amplitude and the number of sidebands will change in accordance with the modulation index[3]. For a faithful reproduction of the input signal, all essential upper and lower sidebands should be recorded. This means that the bandwidth of the recorder for most applications will be an order of magnitude wider than the bandwidths of the data frequency.

In FM recordings the tape transport becomes a possible source of noise. Any speed change, such as flutter, will change the carrier frequency along with resulting sidebands, and these sidebands will show up as noise in the demodulated signal. How much noise will be introduced can be established by the use of the following rule of thumb:

$$\text{Flutter noise} \simeq 20 \log \frac{\text{Deviation in \%}}{\text{Peak Flutter in \%}}$$

If, for instance, a deviation ratio of plus or minus 7½ percent is used and the recorder has a peak-to-peak flutter of 0.15 percent (0.075 percent peak), the signal-to-noise ratio will be limited to 40 db.

A reduction of the flutter-induced noise is possible through flutter compensation. This involves the recording of a control tone, which like the signal, will be frequency-modulated by the flutter. When this control tone upon playback is demodulated and inserted in opposite phase with the demodulated data signal, the noise can be reduced as much as 20 db. (This reduction is limited by the random nature of flutter and skew between the data track and the control track.)

In **pulse-code modulation**, a train of pulses is recorded on the tape. The pulses are again generated by a modulator circuit, where the incoming data signal can vary either the pulse height or the pulse width. If the pulse height is varied, the reproduced signal will be influenced by amplitude variations in the direct recording channel, while pulse-width modulation is affected by flutter and time-base errors.

In **digital** recordings, an analog-to-digital recorder is required. This has the advantage that the data has been recorded in a format that allows for immediate computer processing. This technique is particularly useful in oceanography, where most of the phenomena to be observed varies slowly. It may be necessary to make a recording, say, only once a minute and for this job the **incremental** tape recorder was developed. The analog data is encoded to a digital format which is recorded on the tape. The tape then moves a fraction of an inch waiting for the next recording. These recorders typically operate with a packing density of 256 information bits per linear inch and, consequently, can record for weeks on a single reel of tape, with the advantage that the information can be replayed rapidly. The detailed encoding schemes, digital format, and later computer analysis is more a system problem than a magnetic recording problem; therefore, referred to elsewhere [4, 5].

Seismic data recording requires a tape speed slightly higher than standstill. Stick-slip phenomenon and high flutter make the use of tape inadvisable in the recording of seismic data when the tape must operate at very slow speeds. Because of this disadvantage, the recording media in many seismic recorders is a drum or a disc. Here, the stick-slip phenomenon is eliminated by the inertia of the drum (or disc) and this approach has one advantage: Assuming that data from, say, an underground nuclear explosion has been recorded on a drum with inputs from several transducers spaced far away from each other; it is possible upon later playback to make correlation analysis by moving the individual playback heads with relation to each other.

TELEMETRY RECORDING

While test data from spacecraft and aircraft testing are monitored in real time, they are often recorded simultaneously on magnetic tape. Many industrial companies and laboratories each need a portion of the collected data. The recorded tapes can be duplicated and copies sent to various locations for analysis. In a single spacecraft experiment, numerous transducers and sensors are monitored and in the early phase of using magnetic tape, several channels were recorded to increase the capacity of a recorder. Standards were established very early to make possible the interchange of tapes from various test ranges to firms and laboratories. These standards are compiled in a document called **IRIG 106-66**, which is the latest revision published. They are revised every second year to keep pace with technological advancements. However, no revisions are made that would make obsolete any major portions of equipment and the tape libraries presently in use.

The early standard, which was based upon practical tape recording equipment, called for seven tracks on ½-inch tape or 14 tracks on one-inch wide tape. The number of channels was rapidly exceeded by the number of the transducer outputs to be monitored and multiplex was introduced in telemetry recording. The **multiplex** technique can be based on either a time- or a frequency-sharing basis. Fig. 9-5 illustrates how time multiplex can be accomplished by means of a stepping switch (mechanical or solid-state) and how frequency multiplex is achieved with filters. Time-division multiplex requires that timing signals be transmitted and recorded for synchronous switching to the various demodulators.

In frequency multiplex, FM is often used; therefore, it is referred to as **FM/FM MODULATION.** The IRIG standard specifies the subcarrier channels; i.e., channel number, center frequency, normal deviation, and frequency response. (These standards appear in Chapter 10.) **FM/FM modulation**, in essence, allows 252 data channels to be recorded on a tape transport designed for one-inch tape with a bandwidth of 100 kHz (14 tracks times 18 subcarriers equal to 252 data channels). This technique offers a distinct economical advantage, since it permits analysis of one track at a time using the minimum of 18 demodulators.

In recent years, a newly developed recording technique has been used to capture all modulation sidebands of a telemetry transmission with as little distortion in time and amplitude as possible. This technique is called **pre-detection recording** and it functions as illustrated in Fig. 9-6. The in-

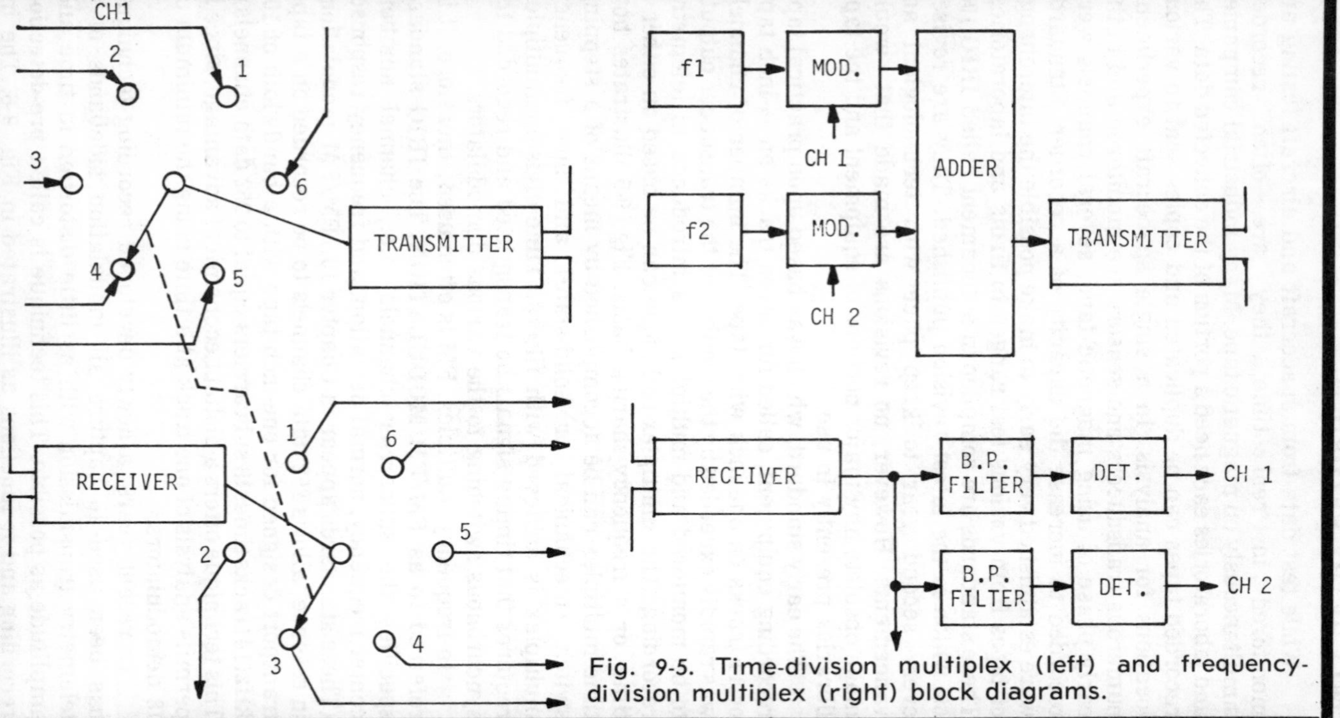

Fig. 9-5. Time-division multiplex (left) and frequency-division multiplex (right) block diagrams.

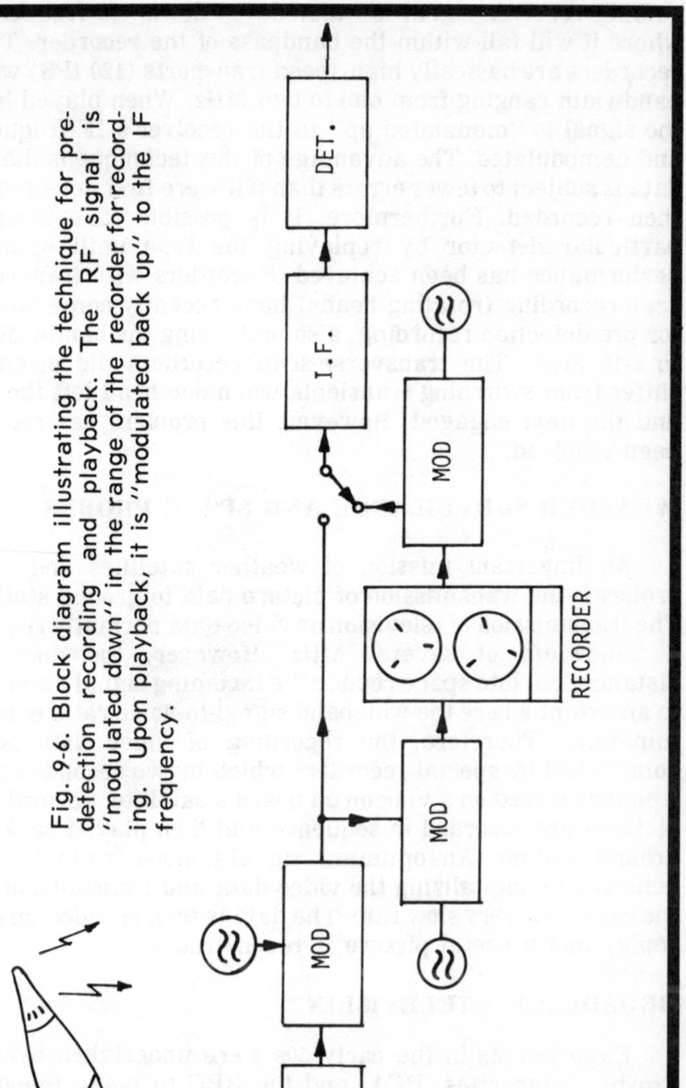

Fig. 9-6. Block diagram illustrating the technique for predetection recording and playback. The RF signal is "modulated down" in the range of the recorder for recording. Upon playback, it is "modulated back up" to the IF frequency.

coming receiver signal is "modulated down" to frequencies where it will fall within the bandpass of the recorder. These recorders are basically high-speed transports (120 IPS) with a bandwidth ranging from one to two MHz. When played back, the signal is "modulated up" to the receiver's IF frequency and demodulated. The advantage of this technique is that the data is subject to fewer errors than if it were first detected and then recorded. Furthermore, it is possible to optimize a particular detector by replaying the tape until optimum performance has been achieved. Recorders with transverse-scan recording (rotating heads) have recently come into use for pre-detection recording, also, extending the bandwidth up to 6-10 MHz. The transverse-scan recorders did originally suffer from switching transients when one head left the tape and the next engaged. However, this problem has recently been resolved.

WEATHER SURVEILANCE AND SPACE PROBES

An important mission of weather satellites and space probes is the transmission of picture data to ground stations. The transmission of television or video data normally requires a bandwidth of several MHz. However, the enormous distances out into space reduce the incoming signal amplitude to an extent where the wideband signal-to-noise ratio is below minimum. Therefore, the recording of video data is accomplished by special recorders which in two seconds record a picture stored on a vidicon on board a satellite. Several such pictures are recorded in sequence and then played back to a ground station. An optimum signal-to-noise ratio is often achieved by digitalizing the video data and transmitting it to the earth at a very slow rate. The data is then speeded up upon replay and a useful picture is reconstructed.

BROADCAST TELEVISION

Experiments in the early 50s were undertaken by Bing Crosby Enterprises, RCA, and the BBC to make television recording possible. In one scheme the required extended bandwidth was obtained by moving the tape at a speed of 20 feet per second, a speed with inherent difficulties because of the air film formed between the magnetic head and tape. Another scheme involved breaking the video signal down into several separate channels by filters. Then, by suitable modulation the signals were to fall within the bandpass of a multi-channel recorder. But due to skew between the individual tracks this method was never successful. The time-base stability for a television picture recording must be less than 30 nanoseconds (one nanosecond is equal to 10^{-9} second). The skew between tracks in a good instrumentation recorder is in the order of 300 nanoseconds, and with the frequency-filtering technique the time-base error was on the order of 500 nanseconds. Therefore, the pictures recorded and played back were of poor quality with regard to stability and freedom from jitter.

Another disadvantage of the high-speed video recorders was the poor low-frequency response. Ampex Corporation, in America, consequently took a different approach to the problem: A two-inch wide tape was (and is) used at a longitudinal tape speed of 15 IPS: the video information is recorded and played back by a rotating assembly, containing four heads, which scans the tape transversely. This principle was shown in Fig. 3-3; it provides for a writing speed of 1550 inches per second. Rather than recording the video information directly, Ampex further decided to use a modified FM modulation technique. This technique employs a much higher deviation rate than normally found in FM systems and the carrier is near the cut-off frequency of the recorder, so only a portion of the carrier and the lower sidebands are recorded and reproduced. (This technique is generally referred to as vestigial sideband recording.)

Recorders using a rotating head assembly in transverse-scan recording and playback require more servo-controls than are normally found in an instrumentation recorder. When playing back a recorded tape, the heads on the rotating assembly must match up with the recorded tracks, and this is done by recording a control track on one edge of the tape. The transition from the time one head leaves the tape until the time the next head engages it must further take place in the proper timing to prevent geometrical distortion of the reproduced picture (venetian blinds). The reproduction of a single picture frame requires several scans and the basic servo mechanism to avoid a geometrical distortion is provided by the guide that wraps the tape against the rotating head assembly. In the last few years Ampex has developed electronic devices for further correction of any geometrical distortion or switching transients and the time-base error is now as low as a few nanoseconds, which is important for the recording and the reproduction of color television programs. The color information in the American NTSC standard is contained in a shift in the phase of the color subcarrier and this phase shift must be within a few degrees for proper color reproduction.

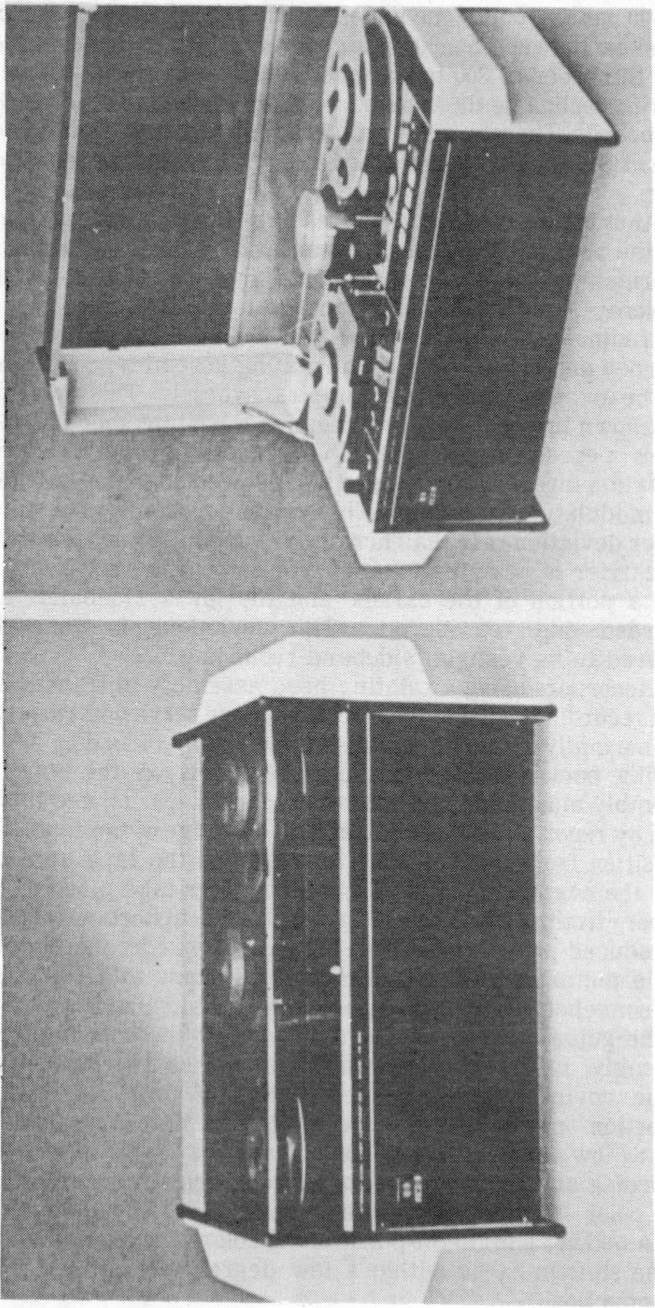

Fig. 9-7. Helical-scan video recorder. (Courtesy I.V.C. Corp., U.S.A.)

EDUCATIONAL AND INDUSTRIAL VIDEO

A broadcast video recorder with color and animation capabilities costs close to $100,000, which obviously is out of the price range for schools and industry and most certainly for home use. The servo mechanisms required for transverse-scan video recording are complex and costly, so in the early 1960s a new design—called helical-scanning—was approached. In a helical-scan recorder the relative speed between the magnetic heads and the tape is again obtained by a rotating head assembly. The tape is moved at a slow speed and is scanned longitudinally at a slight angle; one such machine is shown in Fig.9-7. A single recorded track contains the picture information of one frame; therefore, the helical-scan technique provides the advantage of being able to reproduce or play back a picture while the tape is at a stand-still. Servo-controlled tape motion in the helical-scan recorder is still necessary, in particular with regard to tape tension and edge guidance. A typical standard for a recorded tape is shown in Fig. 9-8, with other standards listed in Fig.9-9. Interchangeability between tapes recorded on different make helical-scan machines is impossible due to the absolute lack of standardization which undoubtedly will delay the growth of this segment of magnetic recording.

MANAGEMENT INFORMATION

The storage capability of the magnetic media has drastically changed bookkeeping information for management. Modern computers are using reel-to-reel type tape mechanisms in addition to magnetic disc files and core memories. A disc file basically consists of a circular disc coated with a magnetic recording material and it is commercially available in a unit called a Disc Pack, as shown in Fig. 9-10.

The application of magnetic recording in the computer industry, including disc files and video files, is more of a system aspect and is beyond the scope of this book, which has been written with the purpose of providing a fundamental understanding of magnetic recording. Several literature references are provided for the reader who wishes to obtain further knowledge of specialized areas and applications.

FUTURE TRENDS

In looking forward, the superiority of magnetic recording techniques is constantly being challenged by the question,

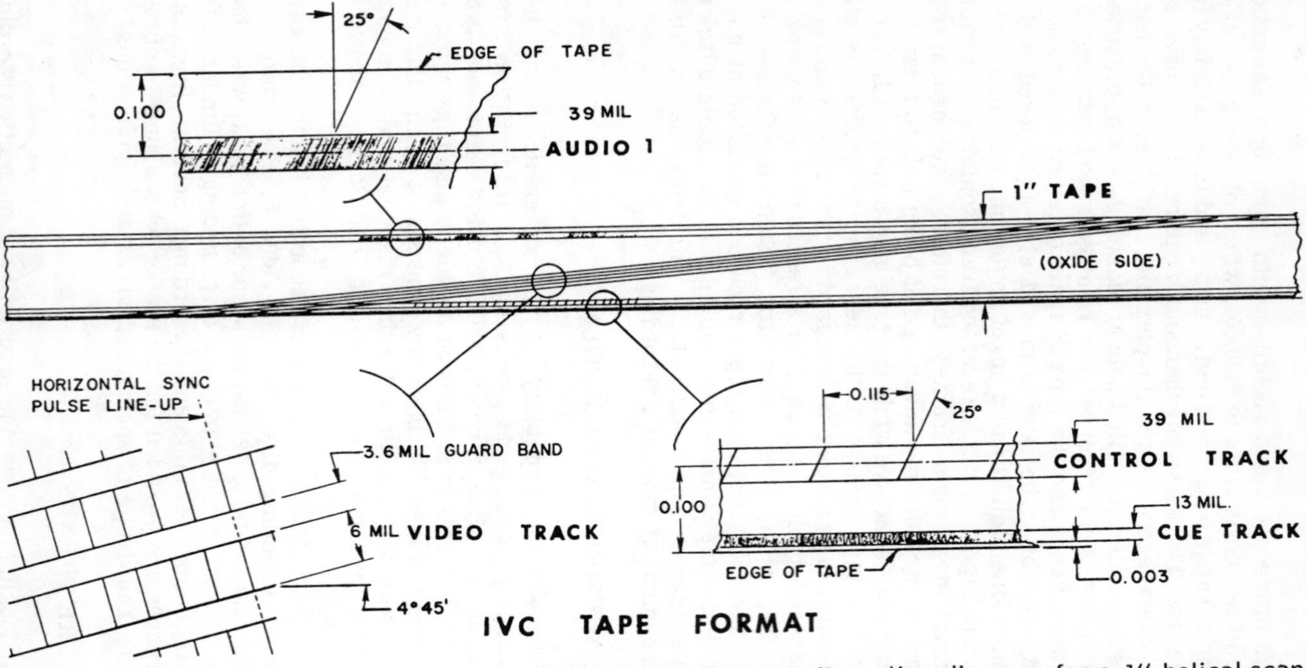

Fig. 9-8. Track configuration diagram for a 1" helical-scan video tape. (Courtesy I.V.C. Corp., U.S.A.)

	IVC 800	Ampex VR660	Ampex VR7000 VR7500	Ampex VR2000	Sony EV200	Sony PV120	PI-3V PI-4V	Norelco EL 3401	Dage DV300	Wollensak VTR150
Tape Speed - ips	6.91	3.7	9.62	15	7.8	4.25	7.5	9.0	5.91	7.5
Writing Speed - ips	723	640.3	1000	1550	590	740	631.5	1086	618	180
Number of Video Heads	1	2	1	4	2	1	2	1	2	1
Wavelength for Peak White (Microinches)	111	98.0	182 154	155	122.9	148	126.3	252.6	127.4	100
Recorded Video Track Length (Inches)	11.94	6.756	16.65	1.8	9.89	12.38	10.41	18.57	10.20	3.13
Video Track Width (Mils)	6	7.5	6	10	6.2	7.1	8	6	5.91	15
Video Guard Band (Mils)	3.6	2.1	2.67	5	3.4	2.2	2	1	1.77	5

color

Fig. 9-9. Existing "standards" for helical-scan video recorders. (Courtesy I.V.C. Corp., U.S.A.)

Fig. 9-10. Disc Pack being loaded into a disc drive. (Courtesy Memorex Corp., U.S.A.)

"What other methods are available?" A mechanical recording device, such as the phonograph record, has been able to hold a stand against magnetic recording in the sound recording industry, and aside from the writing oscillograph and high-speed motion picture camera in industrial and telemetry applications, there have been only a few inroads into the high-frequency recording area. Thermoplastic recording, which requires an electron beam and a vacuum chamber, was introduced and developed by General Electric Company in the late 1950s and laser-beam type recording has also been pursued, but neither method has been very successful. The reason for failure in exploiting these areas lies primarily in the mechanical stability required. Time-base error stability in a magnetic recorder is difficult to achieve in itself and adding to that any temperature-dependent variable, such as the expansion coefficient of different materials, there is obviously a mechanical limitation for the storage capability of a certain media, no matter how theoretically promising it looks.

Areas for progress lie predominantly in the improvements of present magnetic recording techniques. Magnetic heads presently suffer from inefficiencies at high frequencies and techniques are being developed for depositing low-loss ferrite materials with a pole-tip of "Alfesil" or "Alfenol" or "Vacodur," which are trade names for an aluminum-iron alloy with high permeability, low loss and high wear resistance. In magnetic tape manufacture, a departure from standard gamma iron oxide is evident with DuPont's introduction of chromium dioxide. This latest departure can develop into an expansion of the magnetic materials used for magnetic tape. This can be in the form of other new magnetic particles as in a coated tape or in the deposition of a metallic layer on a suitable film or substrate.

References

1. Daniel, E. D.: "A Flux Sensitive Reproducing Head for Magnetic Recording Systems," Proc. IEE (London), Vol. 102, Part B, No. 4, July 1955.

2. International Telementry Conference, Los Angeles, Oct. 1966.

3. Hund, A.: "Frequency Modulation," McGraw-Hill Book Company, Inc., New York, 1942.

4. Hoagland, A. S.: "Digital Magnetic Tape Recording," John Wiley & Sons, 1963.

5. Bycer, B. B.: "Digital Magnetic Tape Recording: Principles and Computer Applications," Hayden Book Co., 1965.

6. Radiation, Inc. Staff: "Digital Techniques in Airborne Data Acquisition," Automatic Control, 37 (February 1961); 41 (March 1961).

7. Thompson, R. S. and L. E. Head: "High Data Capacity/High Environment Recorders," Proc. Intern. Telemetering Conference, Washington, D. C., 625-644 (1965).

8. "Missile Re-Entry Photograph Analyzer with Magnetic Tape Recorded Output," Electronic News (June 22, 1964).

9. Wright, Robert E.: "How to Make Computer Compatible Data Tapes," Control Engineering, 127 (May 1962).

10. Davies, G. L.: "Magnetic Tape Instrumentation," p. 176, New York, McGraw-Hill Book Co., 1961.

11. Johnsom, G. Nels and W. R. Johnson: "A Predetection Recording Telemetry System," IRE Natl. Conv. Record, Pt. 5, 209 (1961).

12. Crosby, J. C. and C. E. Wright: "Effect of Magnetic Tape Recorder Performance on Predetection Signals," Proc. National Telemetering Conference, Los Angeles, 9-1 (1964).

13. Ginsburg, C. P.: "Rotary Head Recording of Wideband Analogue and Digital Data," Proc. Intern. Conf. on Magnetic Recording, London (1964).

14. Schulze, Glen H.: "Applications of a Light Mass Capstan Tape Recorder," Proc. National Telemetering Conference, 9-5 (1962).

15. Himmelstein, S.: "Design of a Multichannel Magnetic Recording System for Frequency Multiplication," IRE Trans. Audio, 9, No. 5, 166-173 (Sept.-Oct. 1961).

16. Fairbanks, G., W. L. Everitt and R. P. Jaeger: "Methods for Time Frequency Compression-Expansion of Speech," IRE Trans. Audio, 2, No. 1, 7-12 (Jan.-Feb. 1954).

17. Marlens, William S.: " Duration and Frequency Alteration," J. Audio Eng. Soc., 14, No. 2, 132-139 (April 1966).

18. Lucas, E. D.: "Niniature Tape Recorders," Control Eng., 11, No. 12, 53 (Dec. 1964).

18. Lucas, E. D.: "Miniature Tape Recorders," Control Eng., 11, No. 12, 53 (Dec. 1964).

19. Kuschnerus, Hans J.: "Tape-Driven Display Shows Order-Picker The Way," Contro. Eng., 12, No. 5, 109 (May 1965).

20. Drewry, H. S. and J. R. Howard: "Taking the Giant Step in Power Plant Automation," ISA J., 53 (July 1964).

21. Coskren, R. J.: "Taped Signal Times Tensile Tester," Control Eng., 9, No. 7, 125 (July 1962).

Chapter 10

Measurements & Standards

Several standards for the recorded flux level on magnetic tapes have been estlished in Europe and in the United States, standards that impose design restrictions on both recording and playback amplifiers. In transforming a record current to a remanent flux on the tape, a certain amount of pre-emphasis is necessary to overcome record head losses. Likewise, in transforming the flux on the tape to a voltage at the playback amplifier input, reproduce head core losses and gap losses affect the overall frequency response. Therefore, it is important that the designer knows not only the head impedances but also has the data that tells him how the head losses vary with frequency. Both core losses and the gap losses are simple to measure, as described below.

MEASUREMENT OF CORE LOSSES

If the flux level through the core in a magnetic head (without losses) is held constant with frequency, the induced voltage will ideally rise 6 db per octave. The constant flux can readily be provided as shown in Fig. 10-1, by a thin, straight wire placed in front of and in parallel with the gap. Some typical curves for instrumentation heads are shown and the departure from the otherwise straight 6 db per octave line is evident.

The constant current is obtained by connecting the wire to a sine-wave generator in series with an induction-free resistance of a value equal to the recommended termination for the generator. Connecting a VTVM across the resistor provides a measure to keep the current constant, since most generators require slight readjustments of the output as the frequency is changed. A VTVM (high impedance) is connected to the head output terminals to measure the induced open-circuit voltage. Losses in the core reduce the induced voltage at high frequencies and these losses are represented by the

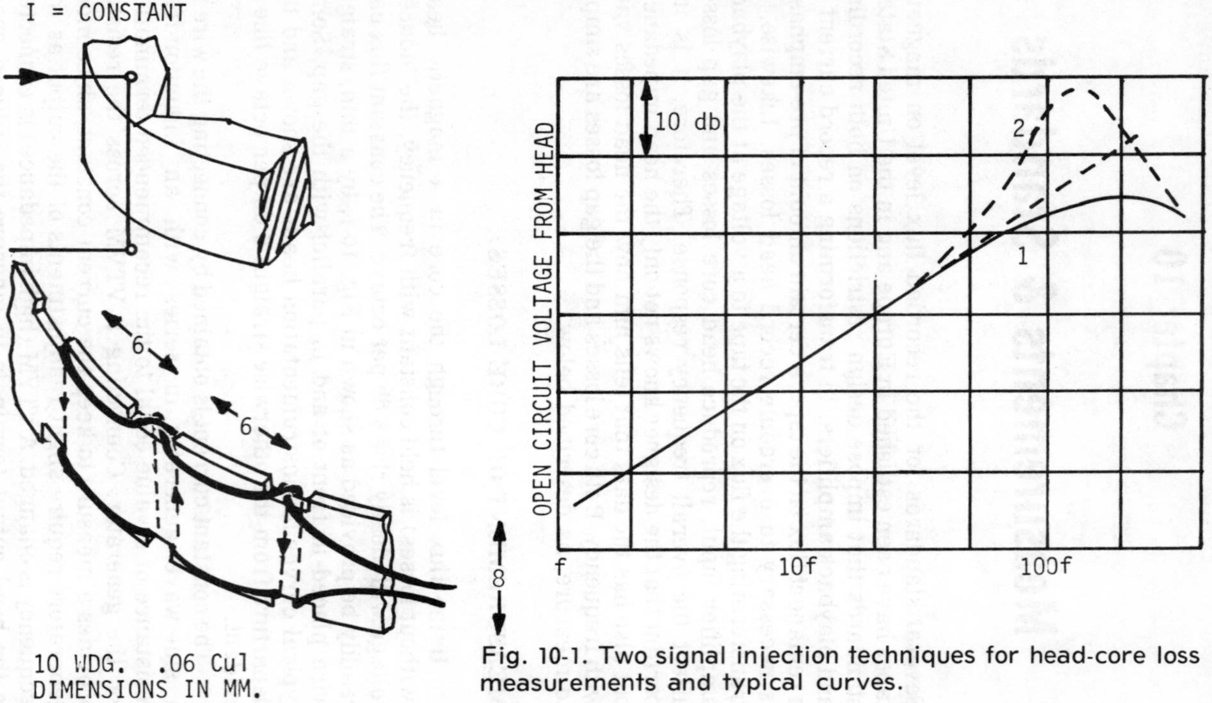

Fig. 10-1. Two signal injection techniques for head-core loss measurements and typical curves.

area between the 6 db per octave line and the measured curve in Fig.10-1. The frequency-response range of the head will also be evident, since the head inductance and its self-capacitance will cause it to resonate at a certain frequency.

Instead of using a straight wire, it may be advantageous to fabricate a small figure-eight loop, as shown in Fig. 10-1. It can be wound on a thin strip of plastic or celuloid, for example, and can easily be positioned in front of the head similar to the path followed by a magnetic tape. The loop should be pressed lightly against the head in such a position that the induced voltage is maximum. Later on in this chapter we show how these curves are applied in the design of pre-emphasis and equalization networks.

MEASUREMENT OF THE EFFECTIVE GAP LENGTHS

Since the magnetic gap lengths are always larger than the mechanical gap lengths, an error can be induced in a design by measuring the gap lengths under a microscope and using this dimension to establish the gap losses. The gap length can be determined accurately only by measuring the wavelength, where the induced voltage from the playback head goes through a null and the gap length calculated from the gap function.

Since this null normally is beyond the frequency range of the recorder (in a properly designed recorder by a factor of 2), the easiest measurement is undertaken by connecting the head leads directly to a tone generator. (The higher frequencies required for this measurement makes the use of high-frequency bias questionable because of the generation of beat notes.) The level of the record current from the sine-wave generator should be of the same magnitude as the bias current normally used, which essentially means the tape is recorded to saturation. A series of tones are recorded and played back; then a curve can be plotted, as shown in Fig. 10-2. The equivalent wavelengths for various tape speeds appear underneath the graph, thus facilitating direct reading of the effective gap lengths. In the example shown, where the tape speed was 3¾ inches per second, the null at 16 kHz corresponds to an effective gap length of 6 microns.

ESTABLISHMENT OF THE PROPER PLAYBACK EQUALIZATION

The simplest and easiest way to adjust a recorder's playback equalization is to use pre-recorded standard tapes. Such tapes are available from several manufacturers for the tape speeds used in audio work. A typical alignment tape has a

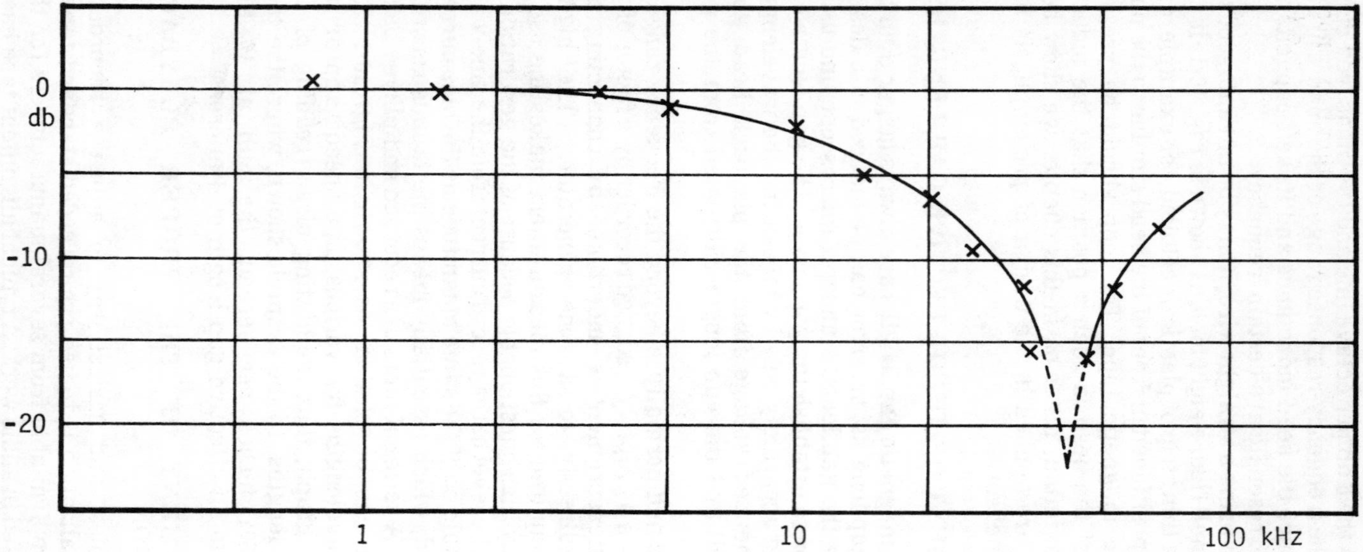

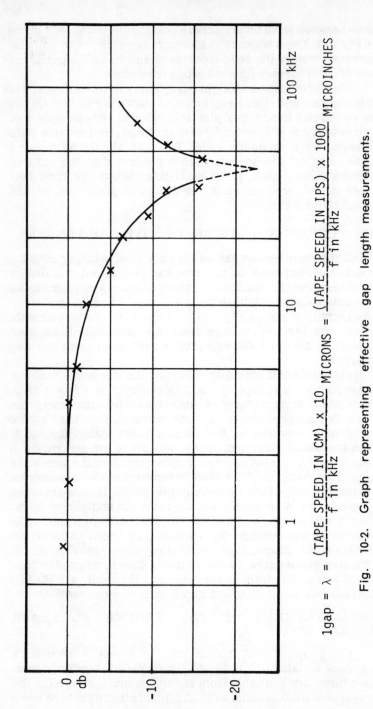

$$1gap = \lambda = \frac{(\text{TAPE SPEED IN CM}) \times 10}{f \text{ in kHz}} \text{ MICRONS} = \frac{(\text{TAPE SPEED IN IPS}) \times 1000}{f \text{ in kHz}} \text{ MICROINCHES}$$

Fig. 10-2. Graph representing effective gap length measurements.

section with a short wavelength recording for proper alignment of the reproduce head. A following section has a normal record level for VU meter zero setting. After these two tones there should be a series of signals, and upon playback of these tones the controls in the equalization circuitry are adjusted for optimum flat frequency response. This procedure will assure that all pre-recorded tapes for a given standard will be reproduced correctly. (Prior to using a test tape, it is advisable to clean the magnetic heads and demagnetize them with a head demagnetizer. It is further advisable not to start and stop the recorder in the middle of a test tone, since this may produce weak transients on the tape and, consequently, shorten its useful life.)

If a test tape is not available, which may be the case with an instrumentation recorder, a different approach must be taken. This technique utilizes the figure-eight loop (or a straight wire in front of the reproduce gap), which is fed with a current that has an amplitude versus frequency response as established below. (The input signal to the reproduce amplifier can also be provided by a voltage generated across a small resistor placed in the ground side of the magnetic head.) It is important that the signal be introduced to the playback amplifier through the reproduce head. If the signal is connected directly to the input stage, the head core losses and self-capacitance will not be included in the measurement and the reading will be incorrect. Assuming, at first, that the current through the wire in front of the head is constant and independent of frequency, the flux through the head will be constant but limited by core losses. This corresponds, at low frequencies, entirely to the conditions existing during playback of a tape, but since the external flux from a tape decreases toward the shorter wavelengths, it is necessary to take this into account. The flux through the head core will be further reduced toward the shorter wavelengths because of the gap loss, which can be determined as described earlier.

Assuming that the equalization has already been adjusted, the output voltage from the playback amplifier will increase toward the high frequencies when the wire in front of the head is fed with a constant current. It is relatively easy to predetermine how much this rise in output voltage should be for proper equalization. Referring to Fig. 10-3, the following steps show how to determine the rise in the output voltage:

1. On a sheet of graph paper, plotted for amplitude versus frequency, draw a straight line representing the desired final frequency characteristic (Curve A in Fig. 10-3).

2. Add to this curve the difference between a constant flux

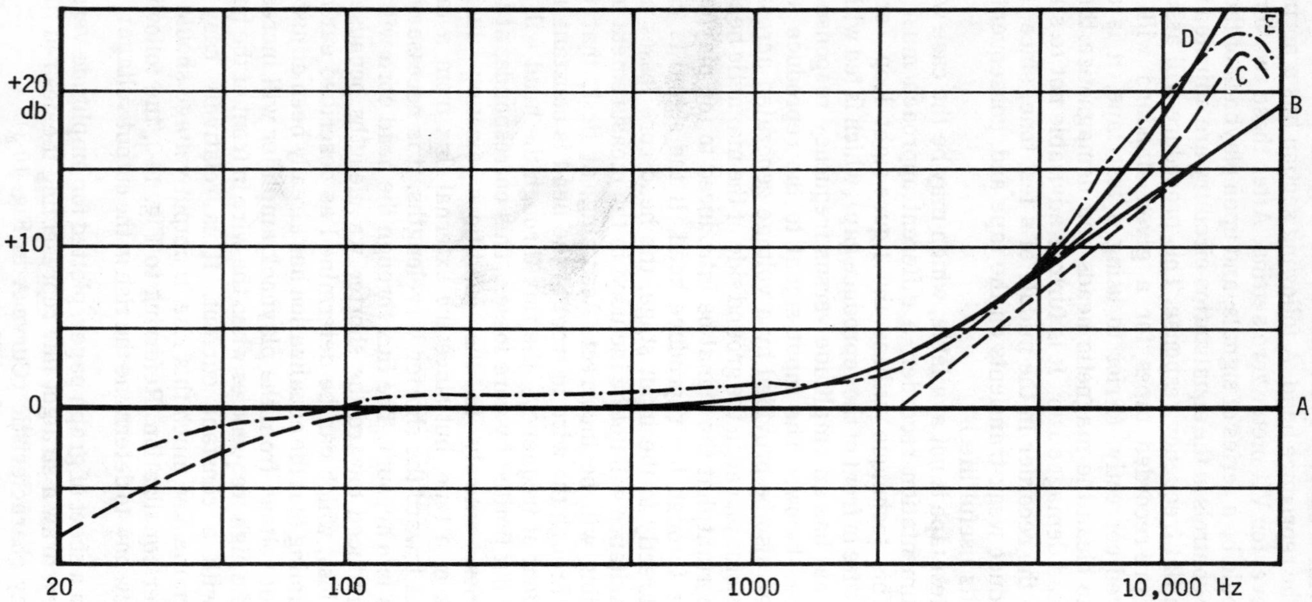

Fig. 10-3. Graph representing post-equalization measurements.

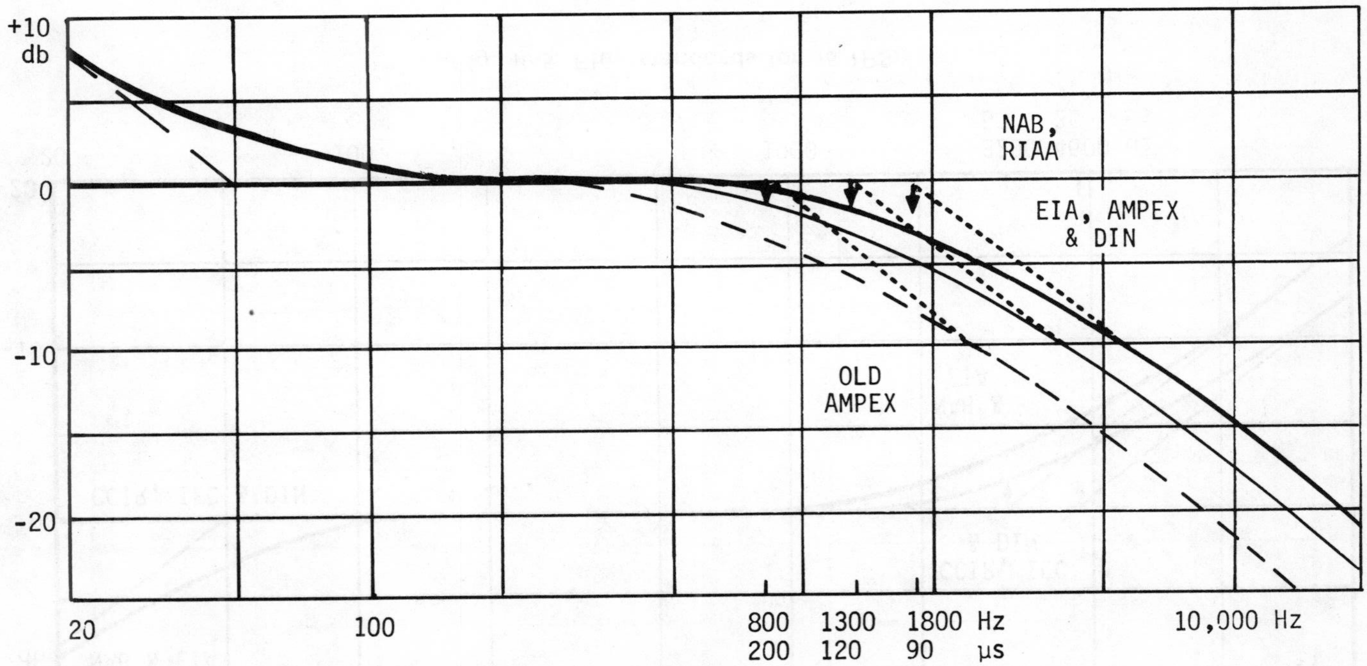

Fig. 10-4. Flux standards for 3¾ IPS.

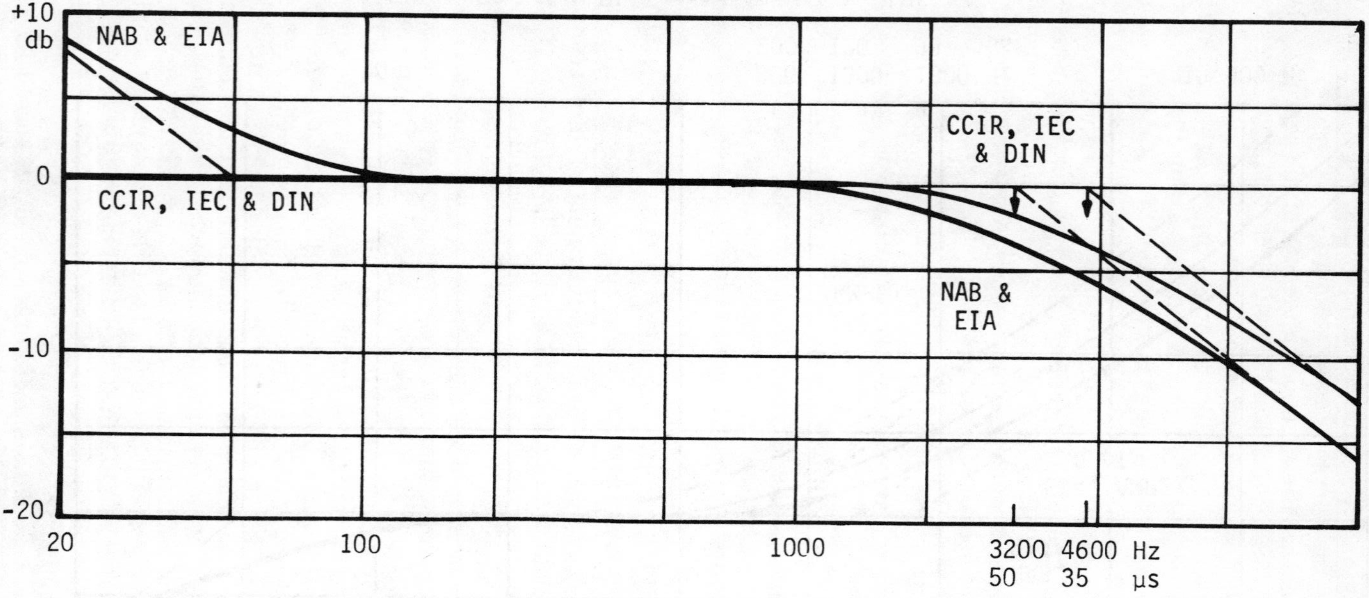

Fig. 10-5. Flux standards for 15 IPS.

recording and a standardized tape flux. Figs. 10-4 and 10-5 illustrate the currently used standards.

3. Measure the effective gap length of the reproduce head as shown in Fig. 10-2. Find the curve for the gap losses from Fig. 4-12. Add these losses to Curve B.

4. Place the straight wire or the figure-eight loop in front of the reproduce head gap in such a position that the output voltage from the amplifier is maximum. If a constant current (versus frequency) is now fed through the wire, the output voltage should closely follow Curve C. An actual measured curve may have a response as shown by Curve D, where the low end will roll off due to amplifier limitations, and it may have a peak at high frequencies due to self-resonance in the reproduce head. If the deviation between Curve D and Curve C is within 1 db, the equalization is as good as can be, but deviations in the order of several db require readjustment (and during design, possibly redesign) of the playback equalizer.

Prior to measuring Curve D, it is advisable to connect an oscilloscope to the playback amplifier and monitor it so that the induced voltage in the playback head does not cause overload and, therefore, an erroneous reading of the output signal.

ADJUSTMENT OF PRE-EQUALIZATION AND THE NORMAL RECORD LEVEL

A prerequisite to proper adjustment of the record amplifier's equalization network and level setting is a properly functioning bias supply. Even small amounts of distortion in the bias supply will generate both noise and distortion on the tape; the latter will affect the proper setting for normal record level (1 percent third-harmonic distortion on the tape).

The best way to check for a proper bias waveform is to energize the record and bias circuitry and place the figure-eight loop in front of the record head with its wires connected to an oscilloscope. The waveform should be a perfect symmetrical sine wave. Most recorders have a 10-ohm resistor in the ground leg of the record head winding which provides a suitable test point to check the bias signal. Many bias oscillators, which also function as erase oscillators, have a balancing potentiometer; an easy way to check for a proper bias waveform is to listen to a tape via the playback amplifier and speaker and to adjust the balancing control for minimum noise. The nature of poor bias or erase noise is a gurgling sound rather than a hiss.

Now, connect a signal generator to the record amplifier

input and observe the waveform across the 10-ohm test resistor in the ground leg of the record-head circuit (or from the figure-eight loop in front of the record head). As a first approximation in setting the correct record level, the **ratio** between the **bias amplitude** and the data **signal amplitude** should be set at approximately 10:1. The tape selected for use on the recorder is now threaded on the tape deck and the tape motion started. In adjusting the proper bias level, there are no set standards, unfortunately, but the two following rules of thumb apply fairly well:

1. For high-quality **audio** work, the bias current is adjusted for 2 db over bias 1 kHz; that is, the bias current is increased from a low value until the recorded signal hits a maximum. The bias current is now increased until the signal has decreased approximately 2 db. This setting normally will assure the largest dynamic range with a minimum of distortion.

2. In **instrumentation** recorders with extended bandwidths, the bias setting is compromised to achieve better short wavelength recording. For example, it is common practice in an instrumentation recorder operating at 120 IPS with a bandwidth of 2 MHz to adjust the bias while recording a 2-MHz tone. The bias level is increased from a low value until the signal on the tape reaches a maximum and then it is increased slightly until the signal level drops approximately 1 db.

The recorder designer, of course, has a choice between these two extreme bias settings and it is up to him to make the decision which necessarily must be a compromise between bandwidth and dynamic range. In all other cases, the reader is referred to the manual for the particular recorder. It should be emphasized again that the bias level is dependent upon the type of tape used and the thickness of its oxide coating, particularly with audio recordings where the higher bias levels are used.

In general practice, what is referred to as the **"normal recording level"** is that level which produces 1 percent harmonic distortion on the tape. With freedom from DC magnetization and a clean bias waveform, this distortion will be a third-harmonic component. The simplified formula for this distortion is:

$$\text{Distortion} \simeq \frac{V3}{V1} \times 100\,\%$$

where V1 is the level of the fundamental and V3 is the level of the third-harmonic component.

Until recently, a frequency of 1 kHz was used in setting up this level for instrumentation as well as for audio recorders. But it will be found that the distortion in wideband instrumentation recorders often increases with frequency. Therefore, it has become acceptable to set the record level in a wideband instrumentation recorder at a frequency that is one-tenth of the upper-frequency limit (for example, 150 kHz for a 1.5-MHz recorder).

While measuring the third harmonic a word of caution is advisable. A sharply-tuned selective VTVM (or wave analyzer) will not measure this component in the presence of wow and flutter from the transport. The bandwidth of the selective VTVM must be fairly wide to capture the third harmonic. The simplest way to do this is to measure V1 with a VTVM, then insert a high-pass filter which cuts off V1 by at least 50 db, and then measure V2. The proper record level is now adjusted by trial and error for 1 percent distortion and the VU meter or other record-level indicating device is adjusted for a reading of 0 db.

The recorder is now ready for adjustment of pre-equalization in the record amplifier. Since pre-equalization in audio work requires a boost of the high frequencies, which cause overload and distortion at the normal record level, the following adjustments should be made at a level at least 10 db below the normal record level. A series of frequencies (for instance, 50, 100, 200, 500, 1000 Hz and 2, 5, 10, 12, 15 kHz) are recorded and played back and the frequency response plotted on an amplitude versus frequency graph. Without pre-equalization it will be found that the frequency response curve will fall off toward higher frequencies and that a corrective network must be installed in a record amplifier or adjustments made to an existing network in order to achieve a flat frequency response. The amount of pre-equalization required is determined from the measured curve and the pre-equalization can be evaluated by measuring the record current across the 10-ohm resistor in the ground leg of the record head with the bias oscillator disabled.

MEASUREMENT OF FREQUENCY RESPONSE

The measurement of a tape recorder's frequency response can be accomplished, as mentioned earlier, by the use of a standard alignment tape, which checks the frequency response of the playback amplifier at the same time. To check the overall frequency response, a signal generator is connected to the record amplifier input terminals and the output is recorded at a number of frequencies and plotted on graph paper. When measuring the response of an audio recorder, it is necessary that the record level be 10 to 20 db lower than the normal record level, since distortion and overload will occur at the higher frequencies, resulting in a frequency response curve showing a roll-off toward high frequencies. This roll-off is unrealistic and is caused by tape saturation due to pre-equalization.

A **sweep generator** is a very useful tool in setting up and aligning a tape recorder. The sweep generator output frequency is swept from a few Hz up to 20 kHz (or more) many times a second; for example, 50 or 60 times per second, and this can be seen on an oscilloscope. This greatly facilitates head alignment and bias and equalizer adjustment. (The single-frequency generator is more useful in the final plot of a recorder's frequency response, though.)

MEASUREMENT OF WOW AND FLUTTER

Several imstruments are available for the measurement of wow and flutter. Actually, a wow and flutter meter is a constant-frequency generator whose output is coupled to the record amplifier while a frequency discriminator demodulates the output from the playback amplifier. The output of the discriminator is fed to a meter which can be reading either RMS or peak-to-peak; the output is likewise available for observation on an oscilloscope. In recorders with separate record and playback heads and electronics, flutter can be measured while the recorder is in the record mode. If flutter is measured in the record mode it will be found that later playback of that tape may give a higher flutter reading due to the flutter occurring during recording playback. It also may be found that some heavy wow components will cancel each other so the playback flutter oscillates between a low value and a high value.

It is important to observe that the signal-to-noise ratio during flutter measurement must be high; in many instances, it is advisable to insert a band-pass filter during playback to eliminate low-frequency and high-frequency noise. Any noise will affect the discriminator circuitry and result in a reading appearing as flutter. Therefore, it is advisable to adjust the record level for saturation of the tape during recording of the particular frequency at which the flutter is being measured.

STANDARDS

Additional material concerning measurement of flutter, time-base error, etc., and existing standards appears in Figs. 10-4 through 10-8. Committees in the United States and Europe are still working on these standards as the state-of-the-art progresses.

Low Band			
Tape Speed ips	± 3-db Pass Band Hz (1)	Record Bias Set Frequency Hz overbias 3-db (2)	Record Level Set Frequency Hz
60	100 - 100,000	100,000 ± 10%	1000 ± 10%
30	100 - 50,000	50,000 ± 10%	1000 ± 10%
15	100 - 25,000	25,000 ± 10%	1000 ± 10%
7-1/2	100 - 12,000	12,000 ± 10%	500 ± 10%
3-3/4	100 - 6,000	6,000 ± 10%	500 ± 10%
1-7/8	100 - 3,000	3,000 ± 10%	500 ± 10%
Intermediate Band			
Tape Speed ips	± 3-db Pass Band Hz (1)	Record Bias Set Frequency Hz overbias 3-db (2)	Record Level Set Frequency Hz
120	300 - 500,000	500,000 ± 10%	1000 ± 10%
60	300 - 250,000	250,000 ± 10%	1000 ± 10%
30	200 - 125,000	125,000 ± 10%	1000 ± 10%
15	100 - 60,000	60,000 ± 10%	1000 ± 10%
7-1/2	100 - 30,000	30,000 ± 10%	500 ± 10%
3-3/4	100 - 15,000	15,000 ± 10%	500 ± 10%
1-7/8	100 - 7,500	7,500 ± 10%	500 ± 10%
Wideband			
Tape Speed ips	± 3-db Pass Band Hz kHz (3)	Record Bias Set Frequency kHz (overbias) (3)	Record Level Set Frequency kHz
120	500 to 1500	1,500	150 ± 10%
60	500 to 750	750	75 ± 10%
30	500 to 375	375	37.5 ± 10%
15	500 to 187	187	18.7 ± 10%
7-1/2	500 to 93	93	9.3 ± 10%
3-3/4	500 to 46	46	4.6 ± 10%

NOTES: (1) Passband response is referenced to the output at the Record Level Set Frequency. (2) Record Bias current is adjusted for maximum reproduce output at a signal level 6-db below Normal Record Level and then increased until an output level 3-db below the maximum value is obtained. (3) Record Bias current is adjusted for maximum reproduce output at a signal level 6-db below Normal Record Level and then increased until an output level 1-db level below the maximum value is obtained.

Fig. 10-6. Direct record parameters. (IRIG Document 106-66)

Tape Speed ips			Carrier Center Frequency Hz	Carrier Deviation Limits		Modulation Frequency Hz	Response at Band Limits db *
Low Band	Intermediate Band	Wide Band		Carrier Plus Deviation Hz	Carrier Minus Deviation Hz		
1-7/8			1,688	2,363	1,012	DC 313	± 1
3-3/4	1-7/8		3,375	4,725	2,025	DC 625	± 1
7-1/2	3-3/4		6,750	9,450	4,050	DC 1,250	± 1
15	7-1/2	3-3/4	13,500	18,900	8,100	DC 2,500	± 1
30	15	7-1/2	27,000	37,800	16,200	DC 5,000	± 1
60	30	15	54,000	75,600	32,400	DC 10,000	± 1
	60	30	108,000	151,200	64,800	DC. 20,000	± 1
	120	60	216,000	302,400	129,600	DC 40,000	± 1
		120	432,000	604,800	259,200	DC 80,000	± 1
Wideband Group 2*							
		120	900,000	1,170,000	630,000	DC 400,000	± 3
		60	450,000	585,000	315,000	DC 200,000	± 3
		30	225,000	292,500	157,500	DC 100,000	± 3
		15	112,500	146,250	78,750	DC 50,000	± 3
		7-1/2	56,250	73,125	39,375	DC 25,000	± 3
		3-3/4	28,125	36,562	19,688	DC 12,500	± 3

* Frequency response referred to 1 kHz output for FM channels 13.5 kHz and above and 100 Hz for channels below 13.5 kHz. The second group of wideband FM carrier frequencies is primarily for use with pre-detection recorders where one or more analog channels are also reouired.

Fig. 10-7. Single-carrier and wideband FM recording parameters. (IRIG Document 106-66)

Channel	Center Frequencies (Hz)	Lower Deviation Limit* (Hz)	Upper Deviation Limit* (Hz)	Nominal Frequency Response (Hz)	Nominal Rise Time (msec)	Maximum Frequency Response** (Hz)	Minimum Rise Time** (msec)
±7.5% CHANNELS							
1	400	370	430	6	58	30	11.7
2	560	518	602	8	42	42	8.33
3	730	675	785	11	32	55	6.40
4	960	888	1,032	14	24	72	4.86
5	1,300	1,202	1,398	20	18	98	3.60
6	1,700	1,572	1,828	25	14	128	2.74
7	2,300	2,127	2,473	35	10	173	2.03
8	3,000	2,775	3,225	45	7.8	225	1.56
9	3,900	3,607	4,193	59	6.0	293	1.20
10	5,400	4,995	5,805	81	4.3	405	.864
11	7,350	6,799	7,901	110	3.2	551	.635
12	10,500	9,712	11,288	160	2.2	788	.444
13	14,500	13,412	15,588	220	1.6	1,088	.322
14	22,000	20,350	23,650	330	1.1	1,650	.212
15	30,000	27,750	32,250	450	.78	2,250	.156
16	40,000	37,000	43,000	600	.58	3,000	.117
17	52,500	48,562	56,438	790	.44	3,938	.089
18	70,000	64,750	75,250	1050	.33	5,250	.067
19	93,000	86,025	99,975	1395	.25	6,975	.050
20	124,000	114,700	133,300	1860	.19	9,300	.038
21	165,000	152,625	177,375	2475	.14	12,375	.029
±15% CHANNELS***							
A	22,000	18,700	25,300	660	.53	3,300	.106
B	30,000	25,500	34,500	900	.39	4,500	.078
C	40,000	34,000	46,000	1200	.29	6,000	.058
D	52,500	44,625	60,375	1575	.22	7,875	.044
E	70,000	59,500	80,500	2100	.17	10,500	.033
F	93,000	79,050	106,950	2790	.13	13,950	.025
G	124,000	105,400	142,600	3720	.09	18,600	.018
H	165,000	140,250	189,750	4950	.07	24,750	.014

* Rounded off to nearest hertz.

** Indicated maximum data frequency response and minimum rise time are based on the maximum theoretical response that can be obtained in a bandwidth between the upper and lower frequency limits specified for the channels.

*** Channels A through H may be used by omitting adjacent lettered and numbered channels. Channels 13 and A may be used together with some increase in adjacent channel interference.

Fig. 10-8. Proportional subcarrier channels. (IRIG Document 106-66)

Appendix

MAGNETIC TAPE MANUFACTURERS

EUROPE

CE

ımel, 12 rue de l'Etoile, Paris 17e
color, 54 avenue de Choisy, Paris 13e

IAN FEDERAL REPUBLIC

-Gevaert AG, 509 Leverkusen, Bayerwerk
ische Anillin & Soda-Fabrik AG, 67 Ludwigshafen,
sota Mining & Manufacturing Co., GmbH, Dusseldorfer
121, 401 Hilden

T BRITAIN

I.I. Tape Ltd., Hayes, Middlesex
. T. Magnetics, (E.V.T. (Bexleyheath) Co., Ltd.),
roadway, Bexleyheath, Kent
rd Ltd., Professional Magnetic Films & Tapes,
203 Wardour St., London W1, (Amateur Zonatape),
Ltd., Ilford, Essex

HERLANDS

ou N.V. Industrieweg, Bladel

U.S.A.

erican Silver Co., Inc., 36-05 Prince, Flushing, N.Y.
pex Corp., 934 Charter, Redwood City, Calif.
pex Magnetic Tape Products, Shamrock Circle,
ika, Ala.
ey Corp., 3500 North Kimball Ave., Chicago, Ill.
lio-Master Corp., 17 East 45th St., New York, N.Y.
rgess Battery Co., Division of Servel, Inc., Foot of
ange, Freeport, Ill.
le Corp., 823 South Wabash Ave., Chicago, Ill.
npco Corp., 1808 North Spaulding Ave., Chicago, Ill.
mputron, Inc., Waltham, Mass.
taphone Corp., 730 Third Ave., New York, N.Y.
gitronics Corp., 100 Albertson Ave., Albertson,
Island, N.Y.
otone Co., Inc., 1 Locust, Keyport, N.J.
ison, Thomas A., Industries, Edison Voicewriter Division,
akeside Ave., West Orange, N.J.
lelitone, Inc., 6415 North Ravenswood Ave., Chicago, Ill.
ebilt Manufacturing Co., Inc., West Pico Boulevard and
nore, Los Angeles, Calif.
oles Engineering, Inc., 247 Greco Ave., Coral Gables,

technical Corp., The, 3401 Shiloh Road, Garland, Tex.
rnational Business Machine Corp., Madison Ave.,
York, N.Y.
ayette Industrial Electronics, Division Lafayette Radio
tronics Corp., 165-08 Liberty Ave., Jamaica, N.Y.
ingston Electronic Corp., 31 Runnymede, Essex Fells,

morex Corp., 1180 Shulman Ave., Santa Clara, Calif.
nesota Mining and Manufacturing Co., 1000 Bush Ave.,
Paul, Minn.
cording Tape Co., 125 East 88th St., New York, N.Y.
eves Soundcraft Corp., 100 Great Pasture Road,
Danbury, Conn.
Rye Sound Corp., Mamaroneck, N.Y.
Sarkes Tarzian, Inc., Bloomington, Ind.
Sweeney Sound, 136 Huron, Toledo, Ohio
Talley Register Corp., 1310-T Mercer, Seattle, Wash.
Teletrosonic Corp., 35-18 37th, Long Island, N.Y.
U.S. Recording Co., 1347 South Capitol, Washington, D.C.

CANADA

Kramer Magnetics Ltd., Port Credit, Ontario
Sound Electronic Specialties Ltd., Scarborough, Ontario

TAPE RECORDER MANUFACTURERS

EUROPE

AUSTRIA

Niwe-Dr. Niedenhuber & Ing., Welzl KG, Am Kunigberg, Vienna 13
Stuzzi Viktor, Ing., Stattemayergasse 30, Vienna 15

BELGIUM

Carpentier Etabits, G. L. S.A. 77 Herelbekestraat, Kuurne (Coutral)

DENMARK

Bang & Olufsen A/S, Struer
Eltra A/S, Maelkevej 3-9, Copenhagen-F
Lyrec A/S, Hollandsvej 12, Lyngby
Movic Denmark A/S, Højnaesvej 56, Copenhagen, Maniøse

FRANCE

Cedamel, 12 rue de l'Etoile, Paris 17e
Crouzet & Cle. Boite Postale 138, Valence sur-Rhone (Drome)

GERMAN FEDERAL REPUBLIC

Beyer Eugen Electrotechnische Fabrik, Theresienstr. 8, 71 Heibronn
Deutsche Philips GmbH, Monkebergstr. 7, 2 Hamburg 1
Elektron Fabrik fur Feinmechanik & Electronik Inh. Ing., Herbert Brause, 6992 Weikersheim
Grundig-Werke GmbH, Kurgartenstrasse, 851 Furth
Haase Karl-Heinz, Schloss-Str. 7, 683 Schwetzingen
Korting Radio Werke GmbH, 8211 Grassau-Chiemgau
Loewe-Opta AG, Posfach 220 & 240, 864 Kronach
Norddeutsche Mende Rundfunk KG, Diedrich-Wilkens-Str. 39-45, 28 Bremen-Hemelingen
Protona Produktionsgesellschaft fur electroakustische Gerate GmbH, Neuer Wall 3, 2 Hamburg 36
SABA-Schwarzwalder Apparatebau-Anstalt Aug. Schwer Sohne GmbH, 773 Villingen
Standard Elektrik Lorenz AG, Schaub-Serk Pforzheim, Oestliche 132, 753 Pforzhelm
Telefunkel GmbH, Ernst-Reuter-Platz, 1 Berlin 10
Telefunkel GmbH, Schulenburger Landstr, 152, 3 Hannover-Hainholz
Tonfunk GmbH, Werdestr. 57, 75 Karlsruhe
Uher Werke MUnchen, P. O. Box 37, 8000 Munchen 47

GERMAN DEMOCRATIC REPUBLIC

Alexander Heinrich, 9935 Markneckeukirchen

GREAT BRITAIN

Ampex Electronics Ltd., Acre rd., Reading, Berkshire
Anglo Netherlands Technical Exchange Ltd., Grosvenor ho. High St., Croyden, Surrey
B.S.R. Ltd., 199 Knightsbridge, London SW7
Elizabethan Ltd., Crow la., Romford, Essex
Fenlow Electronics Ltd., 17 Springfield, la., Weybridge, Surrey
Fi-Cord International, Charlwoods rd., East Grinstead, Sussex
Garrard Engineering Ltd., Newcastle st., Swindon, Wilts
International Computers and Tabulators Ltd., ICT Houseputney, London SW15
Lee Products Ltd., 10/18 Clifton st., London EC2
Leevers-Rich Equipment Ltd., 319b, Trinity rd., London SW18
Magnetic Recording Co., Wyndsor Works, 2 Bellevue rd., London N11
Scopetronics Ltd., 30 London rd., Kingston-upon-Thames, Surrey
Thermionic Products Ltd., Hythe, Southampton, Hants
Vortexion Ltd., 257-263 The Broadway, Wimbledon, London SW19
Wyndsor Recording Co., Ltd., Wyndsor works, 2 Bellevue rd., London N11

ITALY

Brion Vega S.a.s. via Pordenone 8, Milan
Garlanda F.lli Lanificio S.a.s. Fraz Flacero, Valley Mosso
G.F.B. S.p.A. via Gaudenzio Ferrari 3, Turin
Ilotte V. & Crida C. S.n.c. Corso Giulio Cesare 16, Turin
Incis S.n.c. via Novara 28, Saronno
Lesa Construz, Elettromecc, S.p.A. via Bergano 21, Milan
Philco Italiana S.p.A. Piazza Cavour 1, Milan
Smetradio, via S. Antonio da Padova 12, Turin

NETHERLANDS

Amroh N.V. Herengracht 75, Muiden
Hagen W. Dirk Hoogenraadstraat 166, The Hague
Philip's Phonographische Industrie, van Persijnstraat 14, Amersfoort

NORWAY

Proton A/S, Rosenkrantzgaten 11, Oslo
Radionette A/S, Trondheimsvelen 100, Oslo
Tandbergs Radiofabrikk A/S, P.O. Box 411, Oslo

POLAND

Universal, P. O. Box 370, Warsaw 10

SWEDEN

Luxor Industri AB, Motala

SWITZERLAND

Fi-Cord International, 33 avenue du ler-Mars, Neuchatel
Kudelski Stefan, 6 chemin de l'Etang, Paudex, Vaud
Quellet Georges, Ing. elect. E. P. Z. Stellavox, 16 route de Beaumont, Hauterive, Neuchatel
Revox International, 8105 Regensdorf, Zurich

U.S.A.: SOUND RECORDERS

Acton Laboratories, 529 Main, Acton, Mass.
Adeplhi Manufacturing Co., 755 61st, Brooklyn, N.Y.
Allied Recording Products Co., 32-32 Greenpoint Ave., Long Island City, N.Y.
American Printing House for the Blind, 1839 Frankfort Louisville, Ky.
Ampex Corp., 934 Charter, Redwood City, Calif.
Amplifier Corp. of America, 398 Broadway, New York
Antrex Corp., Damen at Willow, Chicago. Ill.
Audio Industries, West 4th & Paul Sts., Michigan City,
Audio Instrument Co., Inc., 135-T West 14th St., New N.Y.
Audio-Master Corp., 17 East 45th St., New York, N.Y.
Aurex Corp., 315 West Adams, Chicago, Ill.
Automatic Radio Manufacturing Co., Inc., 122 Brooklin Boston, Mass.
Autonetics Division, 9150 East Imperial Highway, Down Calif.
Bach-Auricon, Inc., 6950 Romaine, Los Angeles, Cali
Bell & Howell Co., 7100 McCormick Road, Chicago, Ill
Bell Sound Systems, Inc., 6325 Huntley Road, Columbu
Bohn Sound Engineering Co., 6135 Lincoln Highway East Fort Wayne, Ind.
Bridge, Inc., 740 South 42nd, Philadelphia, Pa.
Broadcast Equipment Specialties Corp., Fishkill, N.Y.
Burgess Battery Co., Division of Servel, Inc., Foot of Exchange St., Freeport, Ill.
Code-A-Phone Electronics, Inc., 8136 Southwest Beave Hillsdale Highway, Portland, Ore.
Cohn Electronics, Inc., 5725 Kearny Villa Road, San D Calif.
Criminal Research Products, Inc., Fayette & 2nd Sts., Conshohocken, Pa.
Cross, Herbert & Son, 9-11 Highland Ave., Bala-Cynw Pa.
Dictaphone Corp., 422 Lexington Ave., New York, N.Y
Dresser Electronics, SIE, Division Dresser Industries, Inc., 10201 Westheimer Road, Houston, Tex.
Dynamu, Inc., 110 Duffy Ave., Hicksville, N.Y.
Dynavox Corp., 40-05 21st, Long Island City, N.Y.
Edison, Thomas A., Industries, Edison Voicewriter Division, 31 Lakeside Ave., West Orange, N.J.
Edwards Engineering Corp., 715 Camp, 4th Floor, New Orleans, La.
Electro-Technical Labs, 5134 Glenmount, Houston, Tex
Electromation Co., Venice, Calif.
Electronic Instrument Co., Inc., 33-01 Northern Boulev Long Island City, N.Y.
Electronic Teaching Laboratories, 5034 Wisconsin Ave., N.W., Washington, D.C.
Fairchild Recording Instrument Co., 10-40 45th Ave., Long Island City, N.Y.
Film Crafts Engineering Co., 108 West End Ave., New York, N.Y.
Gables Engineering, Inc., 247 Greco Ave., Coral Gables Fla.
Gamewell Co., 1340 Chestnut, Newton Upper Falls, Mas
General Dynamics/Electronics, Carlson & Humboldt Tr. Rochester, N.Y.
General Electric Co., Defense Electronics Division, Technical Products Operation, Operation Communication Products Department, Electronics Park, Syracuse, N.Y.
Geotechnical Corp., The, 3401 Shiloh Road, Garland, Te
Hohner, M., Inc., Andrews Road, Hicksville, N.Y.
Industrial Electronics, Inc., 127 Light, Baltimore, Md.
International Mutoscope Corp., 44-05 11th, Long Island City, N.Y.
International Radio & Electronic Corp., Department TR, Elkhart, Ind.
Leach Corp., Electronics Division, 717 North Coney Ave Azusa, Calif.
Lekas Manufacturing Co., Ann Arbor, Mich.

etic Recorder & Reproducer Corp., 1533 Cherry,
elphia, Pa.
etic Recording Industries Ltd., 532 Sylvan Ave.,
vood Cliffs, N.J.
er, J. S. Co., 17-08 31st, Long Island City, N.Y.
-Sonics Corp., Fidelity Philadelphia Trust Building,
elphia, Pa.
igan Electronics, Inc., 1744 North Damen Ave.,
go, Ill.
igan Magnetics, Inc., Vermontville, Mich.
estern Instruments, 41st & Sheridan, Tulsa, Okla.
s Reproducer Co., Inc., Department TR, 812 Broadway,
ork, N.Y.
cord Corp. of America, 1915-1/2 Atlantic Ave.,
ic City, N.J.
ronics, Inc., 6514 Woodland Ave., Philadelphia, Pa.
awk Business Machine Corp., 944 Halsey, Brooklyn,

ed Insulation Co., Inc., Baynton & East Price,
elphia, Pa.
cipal Street Sign Co., Inc., 128-04 14th Ave., College
Long Island, N.Y.
comb Audio Products Co., 6824 Lexington Ave.,
ood, Calif.
ronics Co., Inc., The, 8103 West 10th Avenue,
apolis, Minn.
ce Dictation Systems, 5900 North Northwest Highway,
go, Ill.
ron Corp., 777 South Tripp Ave., Chicago, Ill.
more Manufacturing Co., Inc., 130-01 Jamaica Ave.,
ond Hill, N.Y.
e Craft Co., Vanol Building, 427 North Euclid Ave.,
uis, Mo.
ision Instrument Co., Inc., 1011 Commercial,
arlos, Calif.
o Corp. of America, 32 Rockefeller Plaza, New York,

ertone, Inc., 73 Winthrop, Newark, N.J.
ram Corp., 145 East 32nd St., New York, N.Y.
ton Corp., 52-35 Barnet Ave., Long Island City, N.Y.
esound Co., Inc., 35-50 26th, Long Island City, N.Y.
O-Cut, 38-19 108th, Corona, N.Y.
ington Rand Univac, Division of Sperry Rand Corp.,
ast 42nd, New York, N.Y.
re Camera Co., 320 East 21st, Chicago, Ill.
rts Audio Engineering Inc., 216-T Midland Ave.,
River, N.J.
rts Electronics, Inc., 5920 Bowcraft, Los Angeles,

y Sales Co., 2029 North Seventh, Phoenix, Ariz.
, H.H., Inc., 113 Powder Mill Road, Maynard, Mass.
y Recording Instruments Corp., Walter & Green Sts.,
port, Conn.
urg Corp., 2050 North Kolmar Ave., Chicago, Ill.
dscriber Corp., 6 Middletown Ave., North Haven,

S Photo-Cine-Optics, Inc., 602 West 52nd St.,
ork, N.Y.
Products Co., 777 South Tripp Ave., Chicago, Ill.
cil-Hoffman Corp., 845 North Highland Ave.,
ood, Calif.
ny Sound, Inc., 1758 Sylvania Ave., Toledo, Ohio
-A-Phone Co., 5013 North Kedzle Ave., Chicago, Ill.
no Instrument Co., 6666 Lexington Ave., Los Angeles,

ctro Industries Corp., 35-16 37th, Long Island City,

rosonic Corp., 35-18 37th, Long Island City, N.Y.
vision Specialty Division, 1055 Stewart Ave., Garden
N.Y.
po Electronics, Inc., 1341 L N.W., Washington, D.C.
-Ler Radio Corp., 571 West Jackson Boulevard,
go, Ill.

United States Recording Co., 1347 South Capitol, Washington, D.C.
Unity Machine & Tool Corp., Salmon & Westmoreland Sts., Philadelphia, Pa.
Warwick Manufacturing Corp., 7350 North Lehigh, Chicago, Ill.
Webco, Inc., Government Division, 814 North Kedzle Ave., Chicago, Ill.
Webster Electric Co., 1900 Clark, Racine, Wis.
Western Electric & Electric Labs, 805 South 5th, Milwaukee, Wis.
Westrex Co., Division Litton Systems, Inc., 540 West 58th St., New York, N.Y.
Wollensak Optical Co., Division of Revere Camera Co., Ave. D & Hudson Ave., Rochester, N.Y.

U.S.A.: ANALOG RECORDERS

Ampex Corp., 401 Broadway, Redwood City, Calif.
Astro-Science Corp., 9700 Factorial Way, South El Monte, Calif.
Consolidated Electrodynamics Corp., 360 Sierra Madre Villa, Pasadena, Calif.
Hewlett-Packard Co., 1501 Page Mill Road, Palo Alto, Calif.
Himmelstein and Co., South 2500 Estes Ave., Elk Grove Village, Ill.
Honeywell, Inc., Test Instruments Division, 4800 East Dry Creek Road, Denver, Colo.
Kinelogic Corp., 29 South Pasadena Ave., Pasadena, Calif.
Leach Corp., 405 Huntington Drive, San Marino, Calif.
Minnesota Mining and Manufacturing Co., Revere-Mincom Division, 300 South Lewis Road, Camarillo, Calif.
Parsons Electronics Co., Ralph M. The, 151 South Delacey Ave., Pasadena, Calif.
Sangamo Electric Co., Electronic Systems Division, P. O. Box. 359, Springfield, Ill.
Seismograph Service Corp., Seiscor Division, P. O. Box 1590, Tulsa, Okla.
Trak Electronics Co., Inc., 59 Danbury Road, Wilton, Conn.
Winston Research Corp., 6711 South Sepulveda Boulevard, Los Angeles, Calif.

JAPAN

Hayakawa Electric Co., Ltd., 22-22, Nagaike-cho, Abeno-ku, Osaka
Maya Denki Shoji Co., Ltd., Chiyoda Building, 3-22, 1-chome, Koshikawa, Bunkyo-ku, Tokyo
Nippon Electric Co., Ltd., 7-15, 5-chome, Shiba, Minato-ku, Tokyo
Pioneer Electronic Corp., 15-5, 4-chome, Ohmori Nishi, Ohta-ku, Tokyo
Sony Corp., 351, 6-chome, Kita Shinagawa, Shinagawa-ku, Tokyo
TDK Electronics Co., Ltd., 2-14-6, Uchi Kanda, Chiyoda-ku, Tokyo

U.S.A.: DIGITAL RECORDERS

A D Data Systems, Inc., 830 Linden Ave., Rochester, N.Y.
Ampex Corp., 401 Broadway, Redwood City, Calif.
Burroughs Corp., 6071 Second Ave., Detroit, Mich.
California Computer Products, Inc., 305 North Muller St., Anaheim, Calif.
Consolidated Electrodynamics Corp., 360 Sierra Madre Villa, Pasadena, Calif.
Control Data Corp., 8100 34th Ave., South, Minneapolis, Minn.
Cook Electric Co., Data Storage Division, 6401 Oakton Street, Morton Grove, Ill.
Dartex, Division of Tally Corp., 1222 East Pomona St., Santa Ana, Calif.

Datametrics Corp., 8217 Lankershim Boulevard, North Hollywood, Calif.
Datamec, Division of Hewlett-Packard Co., 690 Middlefield Road, Mountain View, Calif.
Digi-Data Corp., 4315 Baltimore Ave., Bladensburg, Md.
Digital Equipment Corp., 146 Main St., Maynard, Mass.
Dymec, A Division of Hewlett-Packard, 395 Page Mill Road, Palo Alto, Calif.
General Electric Co., Information Systems Marketing Operation, 13430 North Black Canyon Highway, Phoenix, Ariz.
GEO Space Corp., Computer Division, 3009 South Post Oak Road, Houston, Tex.
Hersey-Sparling Meter Co., Dacol Divison, 210 West 131 St., Los Angeles, Calif.
Hewlett-Packard Company, 1501 Page Mill Road, Palo Alto, Calif.
Himmelstein and Company, South 2500 Estes Avenue, Elk Grove Village, Ill.
Honeywell Inc., EDP Division, 60 Walnut St., Wellesley Hills, Mass.
Honeywell Inc., Test Instrument Division, 4800 East Dry Creek Road, Denver, Colo.
IBM Corp., Data Processing Division, 112 East Post Road, White Plains, N.Y.
Information Processing Systems, Inc., 200 West 57th St., New York, N.Y.
Kennedy Co., 275 North Halstead St., Pasadena, Calif.
Kinelogic Corp., 29 South Pasadena Ave., Pasadena, Calif.
Leach Corp., 405 Huntington Drive, San Marino, Calif.
Litton Industries, Inc., Data Systems Division, 8000 Woodley Ave., Van Nuys, Calif.
McGraw-Edison Co., Voicewriter Division, 31 Lakeside Ave., West Orange, N.J.
Midwestern Instruments/Telex, 6422 East 41st St., Tulsa, Okla.
Minnesota Mining and Manufacturing Co., Revere-Mincom Division, 300 South Lewis Road, Camarillo, Calif.
Monsanto Co., Data & Control System Department, 800 North Lindbergh Boulevard, St. Louis, Mo.
Oki Electric Industry Co. Ltd., 202 East 44th St., New N.Y.
Parsons Electronics Co., Ralph M. The, 151 South Del Ave., Pasadena, Calif.
Philco-Ford Corp., 3900 Welsh Road, Willow Grove, P
Potter Instrument Co., Inc., 151 Sunnyside Boulevard, Plainview, N.Y.
Radio Corp. of America, Electronic Data Processing Division, Cherry Hill, N.J.
Radio Electronics Corp., 292 Madison Ave., New York N.Y.
Scientific Data Systems, 1649 Seventeenth St., Santa Monica, Calif.
S-I Electronics Inc., 103 Park Ave., Nutley, N.J.
Texas Instruments, Inc., Industrial Products Group, 3609 Buffalo Speedway, Houston, Tex.
Trak Electronics Co., Inc., 59 Danbury Road, Wilton, Conn.
Univac, Division Sperry Rand Corp., P. O. Box 8100, Philadelphia, Pa.
Wyoming Electrodata Corp., 309 East Main, Riverton, Wyoming

JAPAN

Akai Electric Co., Ltd., 883, 3-chome, Kojiya-cho Ota-ku, Tokyo
Eiwa Snagyo Co., Ltd., Rm 308, Kiko-kan Bldg. 3, 1-chome Shiba, Tamura cho, Minato-ku, Tokyo
Munekawa & Co., Ltd., 1-5, 1-chome, Higashi, Ryogo Sumida-ku, Tokyo
Nippon Electric Co., Ltd., EDP Systems Division, 7-1 Shiba Gochome, Minato-ku, Tokyo
Nishikura Industry Co., Ltd., 2, 3-chome, Nihombash Muro-machi, Chuo-ku, Tokyo

Index